ENVIRONMENTAL
INTELLIGENCE
UNIT

FLOODPLAIN MANAGEMENT—
ECOLOGIC AND ECONOMIC PERSPECTIVES

Nancy S. Philippi

The Wetlands Initiative
Chicago, Illinois, U.S.A.

Academic Press

R.G. LANDES COMPANY
AUSTIN

Environmental Intelligence Unit

FLOODPLAIN MANAGEMENT—
ECOLOGIC AND ECONOMIC PERSPECTIVES

R.G. LANDES COMPANY
Austin, Texas, U.S.A.

This book is printed on acid-free paper.

Please address all inquiries to the Publisher:
R.G. Landes Company
909 Pine Street, Georgetown, Texas, U.S.A. 78626
Phone: 512/ 863 7762; FAX: 512/ 863 0081

Academic Press, Inc.
525 B Street, Suite 1900, San Diego, California, U.S.A. 92101-4495

United Kingdom Edition published by Academic Press Limited
24-28 Oval Road, London NW1 7DX, United Kingdom

International Standard Book Number (ISBN): 0-12-554010-8

Printed in the United States of America

Library of Congress Cataloging-in-Publication Data

Philippi, Nancy S., 1935-
Floodplain management : ecologic and economic perspectives / by Nancy S. Philippi
p. cm. — (Environmental intelligence unit)
Includes bibliographical references and index.
ISBN 1-57059-364-7 (alk. paper)
1. Floods—United States. 2. Emergency management—United States. 3. Floodplain management—United States. I. Title. II. Series.
HV609.P45 1996
363.3'4936—dc20

96-20763
CIP

Publisher's Note

R.G. Landes Company publishes six book series: *Medical Intelligence Unit, Molecular Biology Intelligence Unit, Neuroscience Intelligence Unit, Tissue Engineering Intelligence Unit, Biotechnology Intelligence Unit* and *Environmental Intelligence Unit.* The authors of our books are acknowledged leaders in their fields and the topics are unique. Almost without exception, no other similar books exist on these topics.

Our goal is to publish books in important and rapidly changing areas of the biosciences for sophisticated researchers and clinicians. To achieve this goal, we have accelerated our publishing program to conform to the fast pace in which information grows in the biosciences. Most of our books are published within 90 to 120 days of receipt of the manuscript. We would like to thank our readers for their continuing interest and welcome any comments or suggestions they may have for future books.

Shyamali Ghosh
Publications Director
R.G. Landes Company

CONTENTS

PREFACE

Floods are of great interest to the general public during periods of intense flooding, but the ideas of floodplain management that it derives from media reports are overly general and usually confused. Reports of flood damages, particularly, are greatly distorted. Flooding is costly to the general public, however, which funds disaster assistance and flood control projects. These flood control projects, in turn, are costly to the ecologic systems which depend upon the natural flood pulse for their well-being. Floodplains provide a battleground for enmities between ecologic and economic interests, between the general public and the floodplain property owners, between farmers and environmentalists and between federal and local governments. The problem is poorly defined, the issues are misunderstood and the attempts at solutions are uncoordinated. An outpouring of recent publications, in part stimulated by the 1993 floods in the Midwest, have provided good information and ideas, but will not necessarily contribute to more rational floodplain management.

This book clarifies and synthesizes the past development and current status of floodplain management in the United States and points to the directions we must go to resolve our present difficulties. It describes the nature of flooding, its causes and its importance in the ecologic cycle. It defines the conflict of interest between ecologic and economic values inherent in floodplain management. It reviews the historical and logical development of floodplain management policy, driven primarily by private economic interests, in the United States. It describes the relatively recent development of public environmental concerns and attention to ecological values of the floodplain. It identifies the overwhelming problems encountered in the actual execution of agreed-upon floodplain management objectives, and it describes some scenarios, successful and otherwise, that future floodplain management may follow.

At the most practical level, the book presents a synthesis of the volumes of federal reports which have been issued in the first half of this decade, of value to generalist and specialist alike. It provides the historical and logical context in which our present struggle between ecologic and economic values is played out and will offer some new ways of looking at that particular dilemma. This better understanding of the problem points to the areas where solutions lie.

The operative fairy tale of our times is The Emperor Has No Clothes. If the mass media tell us it is so, it must be true; we derive our understanding of floods and their effects from prime time news and, even then, only when it floods. In dry periods the audience for a book on floodplain management is usually limited to the engineers, the environmentalists, urban planners and federal, state and local floodplain managers. Hopefully, however, the 1993 floods have raised the consciousness of floodplain management issues among the general population, since it is the American citizen who cares about both his environment and his pocketbook, in the long run, who would profit most from understanding the information and issues discussed in the following pages.

Nancy S. Philippi

CHAPTER 1

The Challenge of Managing Floodplains

The art of governing has always required compromise between and among the conflicting interests of a citizenry, but there has been none more difficult nor dramatic, in late-20th century America, than the struggle between environmental and economic interests. This struggle, in turn, is no more sharply defined nor fiercely fought than on the floodplains of our continent. Floodplain management, particularly since the early 1970s, has been an unsuccessful exercise in conflict resolution. Our national solution, to date, has been to honor the environmental interests with our words, the economic with our deeds. We have yet, one might say, to put our money where our mouth is.

In the early 1800s it was seen neither as in the public interest nor as a proper function of the federal government to respond to damages caused by floods. Navigation on our major waterways was, however, considered to be an important federal function because it was tied so directly to our economic interests. Our earliest flood control projects were poorly disguised navigation projects. A century later, however, after a series of life and property destroying floods, the Congress adopted the comprehensive precedent-setting legislation, the Flood Control Act of 1936, which has provided the framework for all future federal government interest in floodplain management. The public interest, concerned mainly with loss of public property, was economic.

It was not until the decade of the 1970s, which produced our major pieces of environmental legislation, that ecologic interests became a serious contender in the arenas of public policy, program and allocation of funds. Attention focused on the importance of wetland ecologies to a wide variety of floral and faunal species, providing critical breeding and feeding grounds to some of our most fragile and endangered species. We became aware that, in the interests of preventing flood damages, our programs had destroyed millions of acres of riverine wetland along our major rivers and threatened to destroy many more.

The paradigm is simple. Annual floods are, in most parts of the world, a normal part of the natural water cycle when excessive amounts of water flow out of the river channels and across the floodplains. Floodplains have great economic value. Rich soils from flood deposits, access to river transportation and all the recreational benefits of water make floodplains prime locations for development and for farming. Floods create the economic value here, but they also destroy it, as homes and livelihoods are periodically swept away. Yet floodplains have great ecologic value. Floodwater spreading out across the wide expanses provides feeding and breeding grounds for some of our most unique and interesting wildlife—fish, birds, reptiles and mammals—cleanses the water passing over them and reduces flooding downstream. All this is destroyed when we succeed in keeping floods off floodplains, to protect the economic development. We can manage those floodplains to do one thing or the other but we haven't found a way to do both.

Floodplain management, prior to the emergence of environmental concerns, built structures designed to manage floods: reservoirs to hold the water upstream and release it gradually at a later time; and earthen dykes and levees along the river banks to contain the rising floodwater and prevent it from damaging the crops and development behind them. We then began to recognize the need for nonstructural management tools—programs that would hold the line on economic development, discourage new building behind levees and encourage more natural, ecologic uses of the floodplains. Among them was that crown jewel of nonstructural flood control programs, the National Flood Insurance Program (NFIP), which promised to end flood damages once and for all.

It had become apparent, by 1968 when the NFIP was passed, that the more flood control structures were built, the more damages occurred. Voices, ignored at the time, had warned of such consequences as early as a century before. Protection structures encouraged development behind them which, when overtopped by large floods, resulted in greater damages than would have occurred had those structures not been built. In the 32 years since the passage of our major flood control legislation, the public investment in dams and levees had not prevented flood damages from continuing to climb. It was clear that the federal government would have to play a role in discouraging floodplain development and the NFIP was the tool designed to do the job. In the NFIP the government would make cheap flood insurance available to people if they, in return, would reduce the potential for damage to their homes, and their communities would prohibit future development in the floodplain. The subsidized flood insurance was the carrot; the requirements that local governments manage floodplains and homeowners flood proof and elevate their homes were the sticks; the public investment in subsidized insurance would be rewarded by gradually diminishing flood damages.

As the environmental decade of the 1970s dawned, it promised a constraint, if not a permanent moratorium, upon those economic forces which had been systematically destroying our floodplain habitats. The National Environmental Policy Act (NEPA) demanded that the national planning process be expanded to include, henceforth, an assessment of every public project in terms of environmental costs and damages. Only later would we discover that accounting systems always count the economic benefits of projects far more accurately than they do the environmental costs. The benefits of a stretch of levee designed to protect 100 acres from the 50-year flood can be measured down to the penny, but even if we knew how the ecologic system was affected and just what the losses were, we wouldn't know how to put a monetary value on it. The American public may have put a high value on protecting "environmental quality" but they left the details to the ecologists and, wherever faced with a confrontation between a forest and a farmer, it better understood and generally sided with the latter. Every floodplain management document of the last quarter century—and there have been many of them—has stressed the importance of ecologic preservation and restoration. Of that preservation and restoration, where it floods, very little has been done. Many of the riverine wetlands along our mainstem rivers are no longer wet. And, as we begin the second half of this century's final decade, environmental interests seem to be losing ground. It's not likely that any of us will live to see a presidential campaign whose motto is "it's the environment, stupid." Yet even if lip service to our economic goals moves once again into the ascendancy, floodplains may finally be managed in the most ecologically responsible way—not for the environmental benefits but because, ironically, it makes good economic sense.

The economic costs of satisfying our economic goals have been growing with each passing major flood. Increased protection from floods has meant increased development behind levees, translated into increased damages each time it floods. Our government and our people believe it is a public obligation to reimburse the victims of floods from public coffers. It is a traditional role of government to use public funds to help those individuals who suffer unreasonably at the hands of nature, over which they have no control. These disaster assistance costs have mounted steadily until they reached the neighborhood of eight billion dollars after the 1993 floods. In our money-conscious times the reduction of disaster aid has become an important, if not urgent, public interest.

The NFIP has not, as was predicted, reduced flood damages; nor has anything else in our bag of floodplain management tools. Insurance, disaster aid and structural protection projects all make it more attractive to use floodplains for economic purposes. A phenomenon labeled "moral hazard" by the insurance industry occurs, a fairly common consequence of government programs, whereby helping people who have suffered from an unfortunate circumstance only encourages people to voluntarily seek out that circumstance. The farmer who once knew better than to plant his crops on floodplains will not hesitate to take that chance, today. He might hope for floods, in fact, since federal payments could possibly exceed the market value of his crop. The same motivation might drive a homeowner to build on floodplains, knowing the government would pay for his repairs after the water subsided.

Only when we have learned the economic lessons of the past will we give up spending money on those programs that make it safe and profitable to locate at the rivers' edge, since the end result is only to create greater economic damages. We will instead allow our floodplains to revert back to their natural ecologically-healthy states, because it makes good economic sense to do it.

We may never learn those lessons, of course, and even if we do, they won't be quite so simple. But that's the bare bones outline of the discussion presented in the following pages, with the relevant facts and figures plugged in so that readers can reach their own conclusions. The details of floodplain management become so complicated, sometimes, that it is hard to sort them out and see them clearly. Our management programs have gotten better and better, with time, but even as they have, we have drifted further from our goals. Perhaps it is true that, as Voltaire said and Gilbert White paraphrased in 1966,[1] the better is the enemy of the good.

The following pages attempt to provide a bird's-eye view of the floodplain management field and how it has shifted to include ecologic goals in what began as a purely economic interest. Chapter 2 presents a way of looking at floods as a benign and natural phenomena that, nevertheless, we would do well to avoid. It includes the beginning of what becomes a continuing discussion about the engineering and management techniques of defining and describing floodplains and floods. Chapter 3 examines the reasons that we try to manage floodplains: to protect and preserve the ecologic values and to reduce and eliminate the economic damages. Chapter 4 reviews the history of floodplain management and how it was decided who should do what. Chapter 5 presents the results of those decisions in a description of existing floodplain management programs at the three levels of government, federal, state and local. Chapter 6 examines a problem that we have created in our attempts to solve the problem of flood damages—the damages created by the flood control projects, themselves. Chapters 7 and 8 step back to review where we are today. Chapter 7 discusses our floodplain management policy, as we talk and write about it in official pronouncements. Chapter 8 describes the many problems that have been encountered as we have tried to implement those policies in the real world. Chapter 9 discusses two very interesting documents that were published in 1995, one broad in

1 White said the search for the best in terms of economically efficient solutions was the enemy of the good.[2]

scope, the other narrowly confined. The smaller document, a review of some issues that deterred progress in solving flooding problems in the American River basin in California, sheds more light on the subject in many ways than does the second one, an all encompassing Floodplain Management Assessment that the U. S. Army Corps of Engineers did in the upper Mississippi River valley after the 1993 floods. By pure coincidence, no doubt, these two studies deal with the same river basins that we addressed with our first official flood control bill in 1917. Chapter 9 does not attempt to provide a rounded picture of activity in either basin, but simply uses the two documents to expand the picture of current programmatic and policy issues. Chapter 10 provides a subjective evaluation of the field and offers one observer's view of what the future will bring. This view is that most of what we are doing and are saying is not moving us toward actually achieving either our ecologic or economic goals in floodplain management but that we may just reach them anyway.

It should be noted that coastal floodplains are not part of this review. The hurricanes whose winds and subsequent surges subject the Atlantic coastal areas to so much damage are a very different phenomena from overbank flooding. Our regulatory programs encounter some of the same human behavior problems in trying to prevent development in high risk areas, but in the interests of successful floodplain management we need to hone in on what happens there as precisely as is possible.

There is nothing in the following pages that can be considered the final authority on anything. There are a number of recent documents, discussed at length, that are recommended to the interested reader. *Sharing the Challenge: Floodplain Management into the 21st Century*, the Report of the Interagency Floodplain Management Review Committee to the Administration Floodplain Management Task Force, referred to in these pages and the outside world as the "Galloway Report," probably contains all the suggestions that are being made anywhere today on how to approve our floodplain management programs and policies. Its greatest strength, which is also its greatest weakness, is that it contains something for everyone. The associated document, *Science for Floodplain Management into the 21st Century,* the Preliminary Report of the Scientific Assessment and Strategy Team (SAST), prepared also for the Floodplain Management Task Force and known to plain speakers as the SAST Report, is extremely good reading and, although far more limited in scope than the Galloway Report, may be just as valuable. U.S. Army Corps of Engineers Reports do not make particularly good reading, on the other hand, and the *Floodplain Management Assessment of the Upper Mississippi River and Lower Missouri Rivers and Tributaries* prepared by the Corps in response to the 1993 floods makes up for its lack of clarity with its sheer bulk. Probably there is far more useful information scattered through the five appendices to this study than is communicated in the main report. The complete version of the *Floodplain Management in the United States: An Assessment Report* published by the Federal Emergency Management Agency in 1992 is a storehouse of useful information and should be on the bookshelf of anyone interested in the field until it is out of date, which will probably occur tomorrow. A collection of essays that will never be out of date, although it is now ten years old, is the slim volume of proceedings of *The Flood Control Challenge: Past, Present and Future, a* symposium held on the 50th anniversary of the Flood Control Act of 1936. The comments in those pages of such seasoned practitioners in the field as Gilbert White, Joseph Arnold, Theodore Schad, Martin Reuss, Arthus Maass, Leonard Shabman and others provide valuable insights into the areas of their particular expertise.

The disheartening thing about writing a book on floodplain management is that it has all been said before. The more things change, and certainly our programs and policies have, the more things remain the same. Many voices have spoken out, some of them and particularly Gilbert White's, loudly and clearly and over and over again. We have gotten better and better at what we do but we just aren't meeting our goals. Would we be further away from those goals if we had done nothing? Probably. Should we change those goals? Probably not. Can we learn from our mistakes? We can try.

References

1. Kates RW, Burton I (eds.). Selected Writings of Gilbert White, Vol 1. Chicago: University of Chicago Press, 1986.
2. White G. "Optimal flood damage management: retrospect and prospect. In: Kneese AV, Smith SC, eds. Water Research. Baltimore: Johns Hopkins University Press, 251-269.

CHAPTER 2

The Unmanageability of Floods

Flooding just happens. It is an ordinary and natural event, the unpredictable and uncontrollable result of meteorological forces. In the context of those forces, it is not even very important. Other meteorological events have more severe impacts on our lives. Drought, for example, is longer-lasting, further reaching and potentially more damaging. It has been suggested that "When the true costs of drought are known, drought losses can dwarf the losses from other natural hazards"[1] and recently more than half our federal crop insurance payments compensated drought losses while less than 2% went for flood losses.[2] Lightning causes more deaths, on the average, than do floods. Even in Bangladesh, which experiences some of the world's most murderous floods, cyclones produce 100 times more deaths than does riverine flooding.[3]

The meteorological processes that cause floods are influenced by large-scale atmospheric circulation patterns far greater than the affected areas. The floods that are the localized small scale expressions of these patterns, however, are of enormous interest to us because of their dramatic and merciless impact upon our lives and property. Their particular nature varies around the world, in response to different topographic and climatic conditions, but the way they are viewed everywhere is the same—in terms of the effect they have upon the floodplain and the humans, plants and animals that occupy it.

CAUSES OF FLOODS

Flooding begins with the sun whose energy triggers the activity in the atmosphere. Although the atmosphere extends at least a thousand miles out from the earth, weather itself is limited to the troposphere, which ends ten miles above the surface. Within the troposphere the air is constantly in motion through that most important method of heat transfer in weather processes, convection: the air, heated by the sun, expands and rises. The air is warmest over the equator, being closer to the sun, and because our earth is tilted in its orbit, the air is alternately coolest at first one pole and then the other. These contrasts in temperature cause the winds, as air moves away from cooler, higher pressure to warmer, lower pressure areas.

Air circulates around the globe in large wind belts: the doldrums, the trade winds, the prevailing westerlies and the polar easterlies, all moving in roughly circular fashion as warm air pushes poleward, cold air toward the equator. Four factors influence these wind belts: earth's rotation, centrifugal force, frictional and topographical effects, along with the secondary circulation, which is made up of localized wind patterns associated with special features of the earth's surface. Finally, there is the subtle shift of air masses away from the west-to-east direction in which the planet rotates, the Coriolis force.

The basic pattern of the water cycle—evaporation, as water vapor rises; condensation, as it develops into clouds or fog; and precipitation, as it returns to earth as snow or rain—is profoundly affected by these movements of air. Over the portion of the globe containing the United States, air masses collide at the Polar Front (50-60° latitude) and the polar easterlies bump into the prevailing westerlies, causing waves, bulges and low pressure systems that take on a counterclockwise rotation. These systems developing along the Polar Front are extra tropical cyclones or frontal depressions, more commonly known as storms (Fig. 2.1). Since density contrasts between converging air masses are much greater in spring, greater volumes of precipitable water vapor are available then. Therefore extra tropical cyclones and their associated fronts are most common during the months of the year when conditions are likely to be exacerbated by frozen ground, saturated soils and rapid

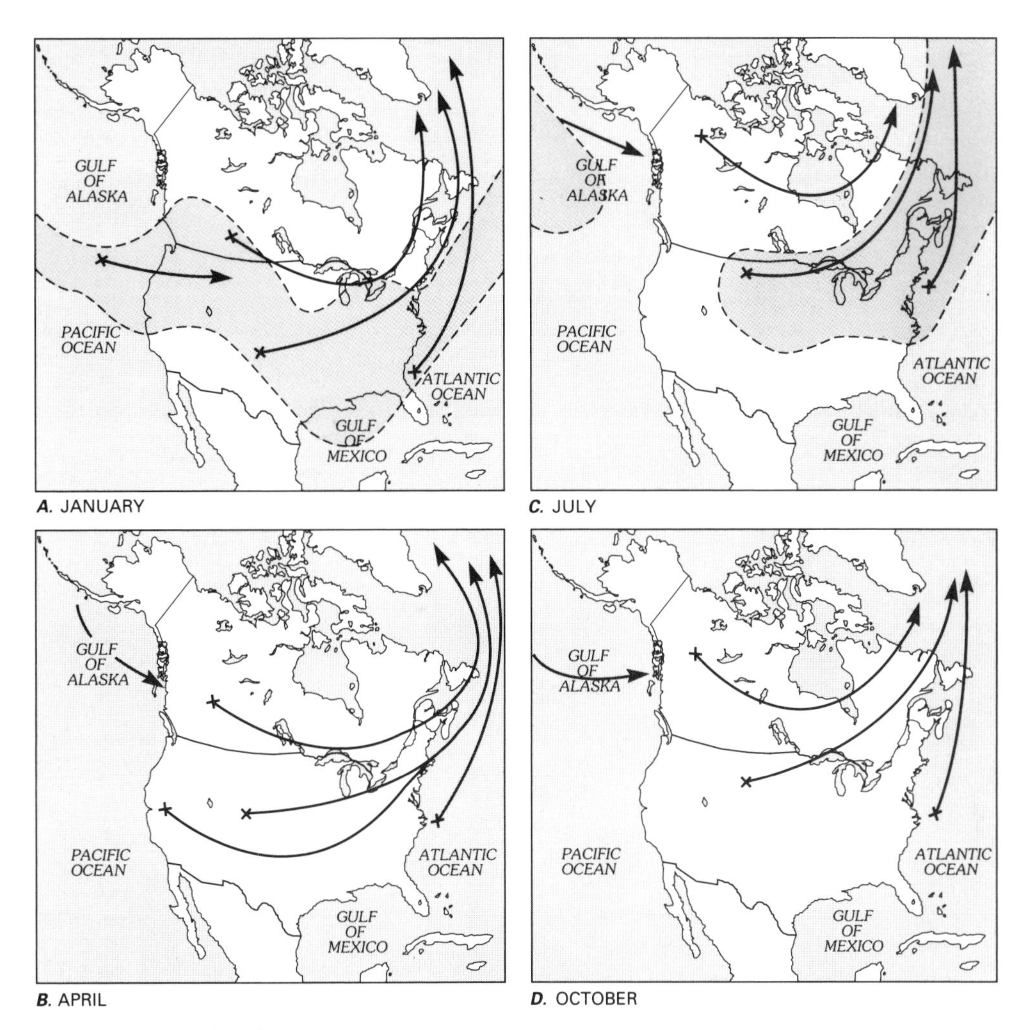

Fig. 2.1. Primary tracks of extratropical cyclones in North America for four midseason months based on frequency of extratropical cyclones during 1951-70. (A) January (B) April (C) July (D) October. Shaded areas indicate areas in which fronts associated with extratropical cyclones occur more than 50% of the time during the winter (December through February) and the summer (June through August). x⟶ indicates primary track of extratropical cyclones. **x** *indicates center of cyclone genesis. Arrow indicates direction. Reproduced from KK Hirschboeck, Climate and Floods, published in National Water Summary 1988-89—Floods and Hydrology, by the US Geological Survey.*

snowmelt (Fig. 2.2). It is these conditions that produce the major floods along the inland rivers of North America (Fig. 2.3).

Flooding of the eastern United States coastline, on the other hand, is usually the product of hurricanes, those violent tropical cyclones which originate in the western Caribbean between August and November of most years. Following a typically northwesterly path which veers, at some point, to the northeast and out to sea, these storms frequently attack the mainland with high velocity winds and ocean surges.

The principal sources of precipitation falling on the United States are the surrounding oceans and the Great Lakes. Moisture is delivered from the Atlantic Ocean and Gulf of Mexico, mainly to the eastern and central United States during the spring and summer; from the Pacific Ocean into North America along several shifting pathways and from the Arctic Region which itself has low moisture content but produces storms when it collides with warmer, moister masses. Moisture is released from the atmosphere by different types of vertical currents that cool and condense moist air: thermal convections or thunderstorms, convergences of contrasting air masses or the forced rising of air hitting the steep slopes of mountain ranges (orographic lifting). Extra tropical cyclones may be stimulated by upper atmosphere wind patterns, as when jet streams overlie atmospheric discontinuities near the surface as they travel from west to east in a wavelike pattern and stall, creating lows like those that produced deluges in the upper Mississippi Basin in 1973 and 1993.

In short, our climate is determined by the way our atmosphere constantly adjusts to the unequal temperatures at the equator and the poles. Climate is best viewed as merely a time average of weather. Weather is unpredictable past a few weeks, after which point the atmosphere "scrambles itself to a point where there is practically no recognition of its initial condition."[4] Climatic conditions, furthermore, may change in response to unpredictable events such as sunspots and volcanic activity. It is as difficult to predict next month's weather as it is the climatic changes of the 21st century.

We do know something of the climate of the past. While systematic observations of climatic phenomena go no father back than,

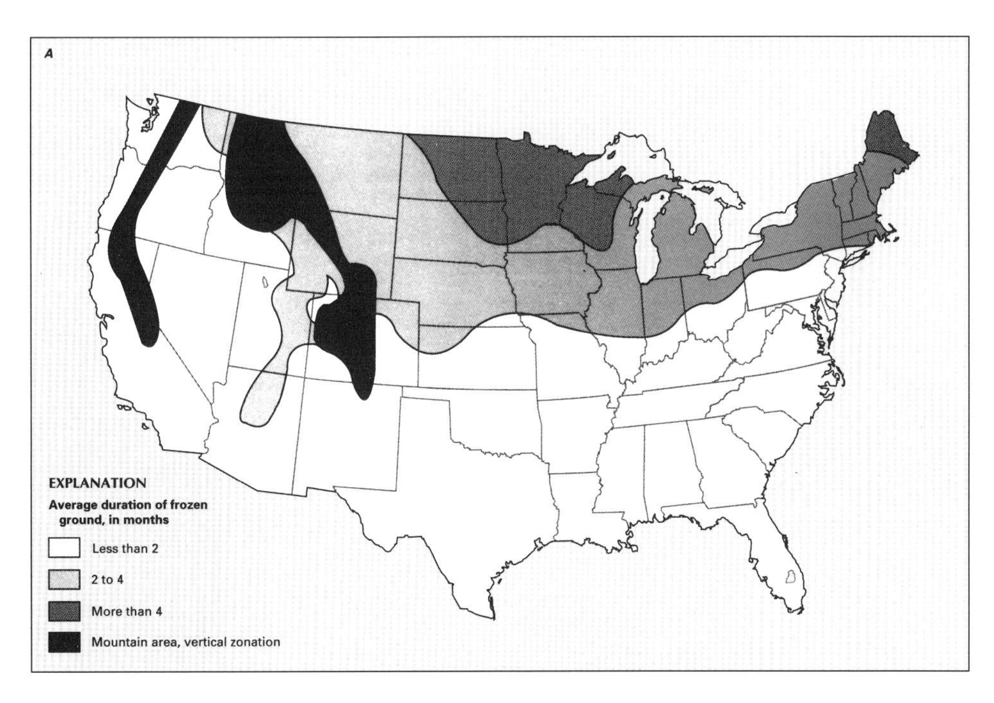

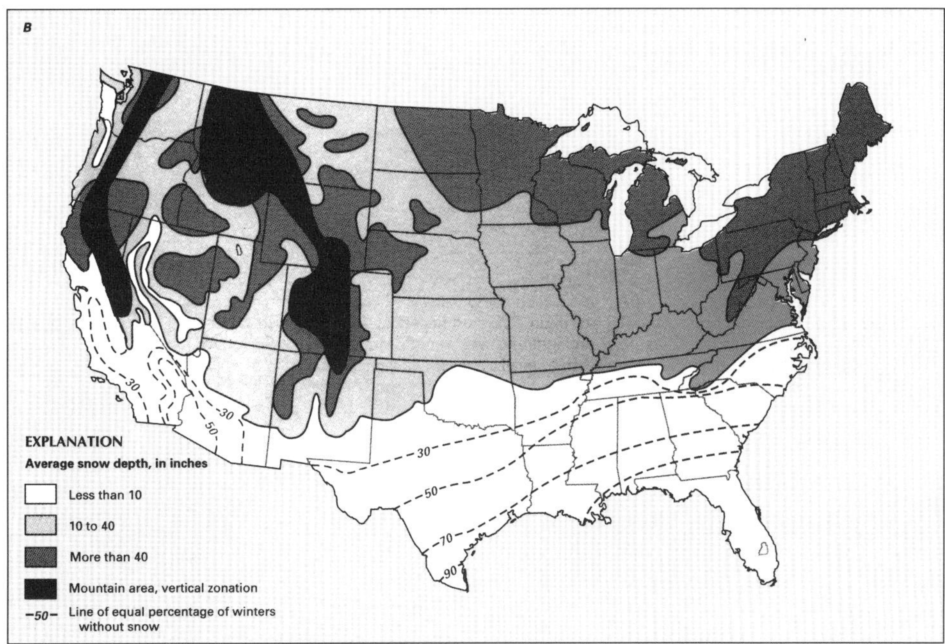

Fig. 2.2. Average duration of frozen ground (A) and average snow depth (B) in the conterminous United States. Reproduced from KK Hirschboeck, Climate and Floods, published in National Water Summary 1988-89–Floods and Hydrology, by the US Geological Survey.

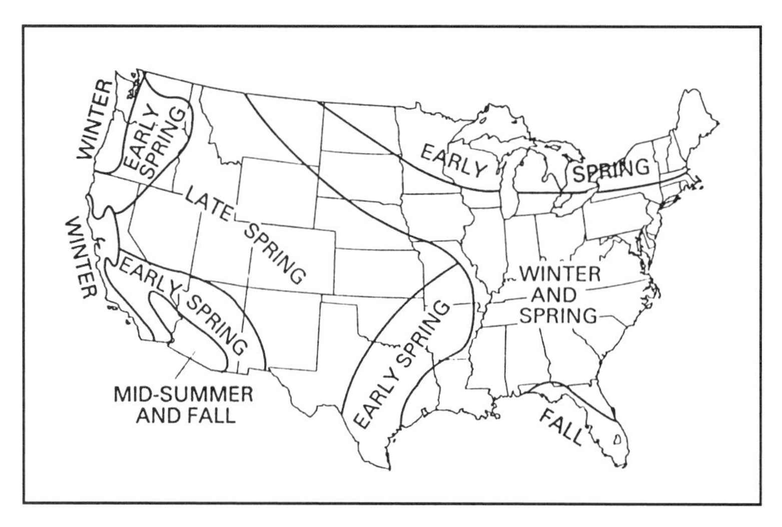

Fig. 2.3. Typical seasons during which the largest flood peak of the year occurs in different parts of the conterminous United States. Reproduced from KK Hirschboeck, Climate and Floods, published in National Water Summary 1988-89–Floods and Hydrology, by the US Geological Survey.

on the average, 35 years,[5] the paleohydrologic record gives us some information on past temperature and precipitation. We know, for example, that the last billion years have been hotter, on average, by 10°C than it is today. In the context of a shorter time frame we are now apparently in an interglacial period whose warming began ten to fourteen thousand years ago. During the last one thousand years the year 1200 AD was the mildest, after which it gradually cooled until 1895, since which date the glaciers have been receding and temperatures rising.[6] The Intergovernmental Panel on Climate Change has observed that global mean surface air temperature has increased by 0.3 to 0.6°C over the last 100 years.[7] In the context of the longer time frame this short-term warming trend is not particularly useful in predicting future temperature trends, much less the floods they could conceivably produce.

Analysis of tree rings tells us that there have not been any long-term changes in mean annual precipitation over the past 400 years, but within that time frame there have been wide swings in precipitation variability over 20- to 30-year periods. The 45 years between 1920-1965, for example, was a period of low variability and low precipitation.[8] As most of our precipitation gages have been providing systematic data, on average, only since 1942, the historical record is probably of no more use in predicting precipitation than it is in predicting temperature trends.

THE NATURE OF FLOODS

Given the climatic patterns to which the United States is subjected, it is normal to experience heavy precipitation and therefore flooding in the spring. According to the United States Geological Survey (USGS), "Floods occur when weather deviates strongly from the long term climate pattern,"[9] but the conditions which lead to them can hardly be described as deviate. One observer has described floods as caused by weather that delivers more precipitation to a drainage basin than it can absorb or store[10] but this is simply a definition of normal precipitation. A drainage basin, after all, is topographically arranged so that the uplands absorb and store and use a portion of the precipitation and then release the remainder into channels, which begin as rivulets and end up as the mainstem rivers that empty into oceans. These channels are cut to depths that accommodate the flows experienced most often. But flow is no more constant than precipitation and in the winter and the spring the waters can be expected to overflow the normal channel banks and spread across another important feature of the watershed, the floodplain. Floodplains are those ubiquitous and familiar features of our landscape that stretch, sometimes, for miles out from the mainstem channel. They have been smoothed and flattened by the countless floods that have spilled out over them, and they bear witness to the regularity with which a river overflows its banks. A truly deviate flood would be the one which extends beyond the natural floodplain, a flood that by definition hasn't happened yet. Flooding, as any bell curve distribution, has its extreme events. When unique combinations of atmospheric moisture, frequency, location and persistence of storms, and seasonal variation of land conditions occur to produce the exceptional floods, those are the ones that never are forgotten. Apocryphal floods, along with the deluge of forty days and forty nights in the Judeo-Christian account, occur in the folklore of American Indians, Fiji Islanders and Australian Aborigines, in Greek mythology and in ancient Babylonian and Mesopotamian legends.

The particular form a flood will take depends upon the landscape on which precipitation falls. The composition of the bedrock; the surficial deposits and the soils; the amount of land cover and the type of vegetation; the topographical relief and channel slope; the temperature and the time of year; all these conditions and more, as well as man's own alterations and manipulations of the natural environment, determine the configuration of a flood. Floods happen everywhere and everywhere they are different.

Floods in Ireland are produced by peat failures, where heavy rainfall triggers peat slides producing flood events. The English are more concerned about coastal flooding: their southeast coast is sinking slowly into the sea at a rate of 30 centimeters each century, storms batter the west and south coasts, and a barrier has been built in the Thames river which protects 2% of the population of England and Wales from the "1000-year flood."[11] Tropical cyclones batter the eastern shores, in the Pacific Basin, of many Asian countries and Australia. In China the Yangtze River has produced flood volumes as great as 200 cubic kilometers in 1871 and is being tamed, allegedly, by the Three Gorges Dam that has required relocating more than a million people in the western provinces. In Bangladesh, where fourteen hundred people died in floods in 1988, 30 million people live and work in the delta of a watershed which lies mainly in four other countries. Northern countries have their special problems. In Canada conventional hydrologic tools are not applicable to the ice jam flooding constantly generated by both the freezes and the thaws, and where snowpack volume governs flood peaks. Debris torrents are a major problem in Japan, Canada and the Alps. "Jokulhlaups," those sudden catastrophic releases of impoundments caused by melting ice, are the inevitable consequences of constantly changing glaciers.

On a somewhat more modest scale, the United States experiences its own variations on the prototypic overbank river flooding which occurs in one place or another in this country hundreds of times each year.[12] Flash floods, produced by intense rainfall over short periods of time in mountainous areas with steeply sloping valleys or by unnatural phenomena such as breaking dams, have been known to occur in every state. These floods are too sudden and intense to allow advance warning and are a frequent cause of loss of life. Alluvial fans like those in our southwestern deserts produce floods filled with debris that are also dangerous. Ice jam flooding occurs in at least thirty-five of our northern states while states which contain coastal plains, particularly on the Atlantic Ocean, experience the storm surges and wave action produced by tropical storms and hurricanes. Less common causes of floods are the mudflows typical of certain areas and the fluctuating levels of our larger lakes.

It is not so much the flood itself as the destruction that it causes to people and their property that captures our attention. In the United States many floods have claimed a place in history, notably for their damaging effects (see Table 2.1). Some river

Table 2.1. Some representative major United States river basin floods

1850	Mississippi River and Sacramento River flooding
1862	Mississippi River and Sacramento River flooding
1865	Mississippi River flooding
1869	Mississippi River flooding
1874	Mississippi River flooding
1881	Sacramento River flooding
1889	Dam breaks at Johnstown, Pennsylvania
1891	Sacramento River flooding
1907	Flooding in Pittsburgh and on the Sacramento River
1912-1913	Mississippi River flooding
1913	Ohio River flooding
1927	Mississippi River flooding
1935	Flooding in New England and the Ohio River Basin
1936	Flooding on the Potomac, Susquehanna and upper Ohio (Pittsburgh) Rivers
1937	Ohio River Basin flooding
1951	Major flooding on the Kansas and lower Missouri Rivers
1965	Mississippi River flooding
1972	Flash flooding and dam failure in Black Hills, South Dakota
1973	Mississippi River flooding
1976	Flash flooding in Big Thompson Canyon, Colorado
1977	Flash flood and dam failures near Johnstown, Pennsylvania
1986	Sacramento and Missouri River flooding
1993	Mississippi and Missouri River flooding

basins stand out more than others. Damaging floods occurred on the Mississippi River in 1849, 1850, 1862, 1865, 1869, 1874, 1912 and 1913. When three more waves of flooding occurred again on the same river in 1927, 20,000 square miles were inundated with floodwater, over 200 people were killed, the homes of 700,000 residents destroyed, and it was called "the greatest natural disaster since the San Francisco earthquake."[13] In 1973 Mississippi River floods set new records for the length of time the river was out of its banks and the 1993 flood on the upper Mississippi which set some new records but had less than 50 deaths associated with it, is now being called the *Great* flood. The closer you are to a flood, apparently, the bigger it looks.

Major floods on the Sacramento River in California in 1850 prompted the city of Sacramento to consider relocating itself on higher ground. It stayed on the floodplain, however, and damaging floods returned in 1862, 1881, 1891, 1907 and 1986. Flooding in the Ohio River valley in 1913 killed more than 400 people and caused $200 million in property losses. More disastrous flooding on that river occurred in 1935, when over 200 lives were lost, and again in 1937. Most years bring news of flooding somewhere in the Ohio River basin.

Johnstown, Pennsylvania experienced a dam failure in 1889 that killed two thousand people, more devastating flooding in 1936 and again in 1977. Flooding did terrible damage in Pittsburgh in 1907. In 1935 floods ravaged states from Washington to New York and from Texas north to Wisconsin. In 1936 there was flooding all along the eastern seaboard, over 100 people drowned across the country and floodwater crept out from the Potomac River and across the parks toward the nation's building where lawmakers were crafting the first national flood control legislation. We do not know, each spring, exactly where the flooding will occur, but we do know that it will.

DESCRIBING FLOODS

The record of our early floods is only adequate for anecdotal purposes. The widespread systematic recording of reliable flood data is very recent, and it is only within this record that we can systematically evaluate those characteristics of floods that are of the most interest to us: the volume of water flowing past a particular location in cubic feet per second (discharge), the period of time during which this volume was sustained (duration), and the elevation the water reached (stage). Since each measurement is tied to one single gaging location, our ability to accurately describe a total flood depends upon the number and the placement of measuring devices in that watershed. The USGS maintains a large and important system of recording gages on the nation's more important watersheds: 7,363 were in operation in 1990.[14] Yet of the over 800,000 tributaries in the United States with drainage areas between one and two square miles, fewer than 60 were gaged,[15] and the total number of gages has actually declined since 1980 as a result of cutting back by local governments, who cosponsor the program.

The hydrologic characteristics of a flood that interest us are far more extensive than those recorded by a stream gage. They include, in addition to peak discharges, flood stage and duration, the time of travel and rate of rise of the crests; water velocities; sedimentation and degradation of flood channels and floodplains; the effect of geomorphology on floods and vice versa; hydraulics of flood channels, floodplains and manmade structures; and the effects of floods on water quality.[16] Since we cannot collect such data for every point in every watershed of interest, we use hydrologic modeling to build on the data that we have.

The single most interesting characteristic of a flood is just how big it is. But "bigness" in terms of actual volume or peak stages is meaningless outside a single watershed, and a more sophisticated measure of a flood's size has been adopted: exceedence probability. By analyzing the historical record of floods within a given drainage basin and recording how often floods of various magnitudes have occurred, a curve is generated that describes the probability in any single year of a flood of any given discharge being exceeded. The more often a flood that produced a particular peak discharge has occurred, the more likely (probable) it is to occur again. Floods are described, therefore, in terms of the probability of that particular discharge being exceeded in that watershed in any given year, such as having a 1%, a 2% or a 10% probability. Since this is an awkward way for the layman to think about floods, the terminology has been taken a step further, to describe the probable frequency of reoccurrence. If, for example, a flood has a 1% probability of being exceeded in a particular year, it can be expected, throughout infinity, to happen once in every 100 years. Likewise a 2% flood theoretically will happen every 50 years, a 10% probability flood every 10 years, etc. This terminology can be deceptive though, and is frequently misused because, in the same sense that a flipped coin, which has a 50% probability of producing a head, will often produce tails two times in a row, "100-year" floods occur back-to-back as well. There are many other problems with these calculations, which will be discussed in chapter 6. If, as has been claimed,[17] the typical river reaches the upper limits of its channel every 1.5 years, the probability of any flooding at all on a river is 66%.

Although floods are commonly named in terms of probability or reoccurrence frequency, it is really the peak discharge that is being described. Either or both terms can be used to describe other parameters of a flood and may refer to any area within the watershed. The same meteorological event may produce 100-year discharges on one tributary but not on an adjoining one, in one small portion of a basin but nowhere else. The 1993 flood on the upper Mississippi River was sometimes called a "500-year" event, but only a few stream gages recorded peaks of that magnitude and of the 154 gages in the basin that recorded peaks that rose above the 10-year flood level during that event, half of them recorded elevations below those associated with a 100-year flood event.

This terminology is used to describe historical floods, to predict future floods and to set design standards for flood prevention structures. If an engineer wants to design a levee high enough to prevent the 100-year flood from overtopping it, that levee is said to have been designed to the 100-year level. Floodplains are also described in terms of exceedence reoccurrence.

DESCRIBING FLOODPLAINS

The floodplain is an easily recognizable topographical feature that provides the transition between a watershed's uplands and its stream and river channels. Typically broad and flat, rich in sediments carried from the upper watershed and exhibiting a broad diversity of vegetation, wildlife, fishes and migratory birds, the actual geomorphology of floodplains is affected by the nature of the underlying geologic formations and soils, and by the vegetation, slope and drainage characteristics of the watershed itself. Sometimes there will be no sharp distinction between the floodplain and its river channel as the river meanders back and forth across its valley over time, cutting, sculpting, scouring and filling as it goes. A floodplain such as the one depicted in Figure 2.4 can become extremely complex.

The infinite variations in the annual climatic patterns produce river systems that are always changing. The boundaries of a floodplain are in a constant state of flux, which makes it important ecologically. The floodplain environment shares with wetlands everywhere those special conditions of alternating wetness and dryness so conducive to a diversity and abundance of floral and faunal life. It is that fluctuating boundary that defines the ecologic floodplain.

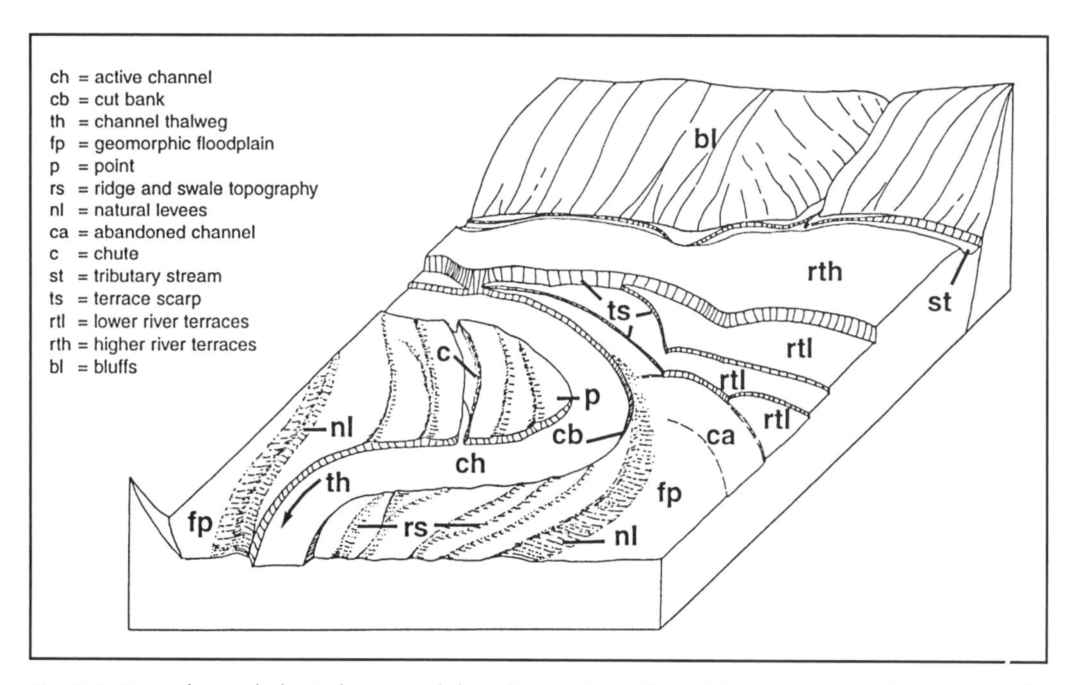

Fig. 2.4. Natural morphologic features of the Missouri River floodplain. Reproduced from Science for Floodplain Management into the 21st Century, Preliminary Report of the Scientific Assessment and Strategy Team. Report of the Interagency Floodplain Management Review Committee to the Administration Floodplain Management Task Force.

The definition of a floodplain for regulatory purposes is quite a different matter. Enforcement of building or development activities upon a floodplain must be equitable and to be equitable, the boundaries are expected to be precise. Enforcing officers must be able to point to a line beyond which the regulation does or does not apply and for this reason the probability or frequency reoccurrence principle is applied. The National Flood Insurance Program (NFIP) has adopted the 100-year floodplain as its regulatory unit, meaning that the program has selected this size flood, the one that has a 1% probability of occurring each year, to define the floodplain within which new development should be regulated.

The U.S. Army Corps of Engineers (Corps) has often designed protection structures to the level of the "standard project flood." The standard project flood is defined as the largest flood that could reasonably be expected to occur at a given location and is generally assumed to approximate the 500-year flood.[18] Agricultural levees that provide flood protection, on the other hand, typically do not keep out floods larger than the 50-year or sometimes only 10-year flood.

The different definitions of floodplains make it hard to estimate amounts of flood prone land. The United States Department of Agriculture (USDA) calculates that there are 1.94 billion acres of surface area in the continental United States excluding Alaska.[19] In 1977 the Water Resources Council (WRC) estimated that 7% or 178.8 million acres of the country, which it defined as including Alaska, Hawaii and the Caribbean, was within the 100-year or NFIP regulatory floodplain. Then in 1987 the Federal Emergency Management Agency (FEMA) estimated[20] that 94 million acres within the NFIP communities were flood prone, which left 84 million acres of 100-year floodplain outside those communities, if the WRC figures were correct. In 1994 the most recent update of our Unified National Plan for Floodplain Management[21] used the 94 million acre figure to describe floodplain area in the United States.

Total floodplain area, however, has been defined to be much larger. In 1982, using data from the early 1970s, the National Resources Inventory (NRI) prepared by the USDA estimated that

195 million acres of rural nonfederal land excluding Alaska were flood prone. Ten years later in their 1992 update of the NRI, (unpublished, provided by the USDA Natural Resources Conservation Service, February 1996) the Department determined that 154 million acres of nonfederal rural land fall into one of three categories of flooding: 54 million acres flood frequently, 53 million acres occasionally and 47 million rarely. The Department had revised their methodology since the 1982 NRI. They are now using evidence related to soils and vegetation to identify lands that fall into the three classes of flooding frequency. The USDA defines those classes as follows:

- Frequent flooding is likely to occur often; under usual weather conditions, more than a 50% chance of flooding in any year; more than 50 times in 100 years.
- Occasional flooding is expected infrequently under usual weather conditions; a 5 to 50% chance of flooding in any year or 5 to 50 times in 100 years.
- Rare flooding is unlikely but possible under unusual weather conditions; 0 to 5% chance of flooding in any year, near 0 to 5 times in 100 years.

According to this definition the 100-year floodplain used by the NFIP to describe floodplains in populated areas falls somewhere in the middle of the category of rural land that "rarely" floods.

Differing definitions of floodplains make it impossible to generate facile conclusions as to size and proportion of floodplain land, but perhaps such generalized numbers are neither appropriate nor useful anyway. It is, after all, quite artificial to impose a rigid construct on a phenomena that by its very definition is constantly changing, and our management practices affect the differently defined floodplains in different ways. Consider that when a levee is constructed along the channel of a river, the effect it has on the floodplain depends on which definition is being used. The topographical feature remains unchanged, a level expanse that stretches smoothly from the channel to perhaps a distant line of bluffs. The ecologic systems that thrived on the transitional wet/dry nature of the floodplain have been destroyed because the transitional zone has been replaced with cropland. The regulatory floodplain that had existed there has either changed or disappeared. The NFIP considers land behind the protection of a levee to be flood free, at least up to the level of protection. A house that once stood well within the 100-year floodplain for example, before the levee was built, is considered to no longer be in the regulatory floodplain. Its foundation may be well below the elevation of the 100-year flood, however, and if the levee is breached or overtopped by a slightly larger flood the damage to properties behind the levee would be devastating.

For these and other reasons, many of them discussed in chapter 6, the definition of the regulatory floodplain as the 100-year floodplain is problematic. From a floodplain management point of view we are concerned about the people and property that are "at risk" of flood damage. Risk is a subjective concept, however, and both governments and individuals make subjective choices as to how much risk they are willing to take. The more catastrophic the consequences, the more difficult it is to absorb risk. The same individual might put his fishing shack in the 100-year floodplain, his home in the 500-year floodplain and his castle on the bluff. Government decisions, likewise, have to be driven by how large the bill for damages will be; by how much property is at risk.

Unfortunately we don't collect the data systematically and so we do not know how much property or how many people are at risk in the floodplains of varying definitions. The 1987 FEMA study that estimated 94 million acres of floodplain derived it from an examination of the 17,466 communities then in the flood insurance program.[22] It estimated that this 100-year floodplain area was used by 9.6 million floodplain households whose property was valued at $390 billion. As recently as 1995 the NFIP administrator estimated that the three million policies in the 18,408 communities in the program represented 25% of the total households at risk (personal communication, September 14, 1995), implying a total floodplain use of 12 million households. If these numbers are comparable then there has been a 25% increase in households on the 100-year floodplain since 1987. (It is unlikely that the increase of 5% in the number of communities in the NFIP would account for more than a small portion of this increase.) In 1978 another study by Jack Schaeffer[23] had estimated that 4.5 million housing units were in "special flood hazard areas," along with an additional 325,000 nonresidential units. If the number of households in the 100-year floodplain really doubled from 1978 to 1987, we need to know it, but without systematic data collection we never will.

The USDA recorded the use of rural floodplains, in the 1992 NRI, and if they continue to use the same methodology in the collection and definition of their data, we will someday be able to describe the trends that we can only guess at for the present. The 1992 NRI acres of rarely to frequently flooded nonfederal rural lands (excluding Alaska) are divided according to the land uses indicated in Table 2.2.

In discussing the use of floodplains, the 1992 Assessment Report references Constance Hunt's estimate in the September-October 1985 issue of the National Wetlands Newsletter, that 35 million acres of indigenous woody riparian habitat remain "in near natural condition" in the lower 48 states.[24] This is reasonably consistent with the NRI numbers.

The land that floods "frequently" in the 1992 NRI is 47% cropland, 13% rangeland and 11% pasture, with the remaining 15% falling into the "other" category, which includes structures, windbreaks, barren land and land in the Conservation Reserve Program.

In summary, flooding is most commonly understood to be the escape of flow from the banks of the river channel. While this occurs, on average, every 1.5 years "there is no readily available estimate of the actual number of floods of a particular magnitude

Table 2.2. Uses put to rural, nonfederal floodplains

1992 Cover/ Land Use	Acreage Flooded Rarely, Occasionally or Frequently	Percentage
Cropland	49,943,000	32.3
Pasture and rangeland	48,528,300	31.4
Forest land	43,757,000	28.3
Other rural land	12,453,100	8.0
Total	154,681,400	100.0

or return frequency that occur in any given year."[25] From a climatic point of view, flood-producing precipitation is a natural and inevitable occurrence and from an ecological viewpoint, the flooding is both desirable and necessary. The problem is, that from an economic point of view, floods cause damage.

There is a wide range in the types of floods and the damages they cause, around the world. Floods in this country are normally described in terms of their size, and their size is defined by the frequency with which they are expected to re-occur. The larger floods occur rarely (i.e., every 500 years) and have a low probability (0.2%) of re-occurring. Smaller floods occur frequently (as often as, e.g., every 10 years) and have a high probability (10%) of re-occurring.

Floodplains are defined by the floods that spread across them. The NFIP has selected somewhat arbitrarily, it could be argued, the 100-year flood to define the floodplain for our most important regulatory program. This means the 1% probability of flood damage risk is the one we try to protect against when limits are put on new development. Our National Flood Insurance Program estimates that twelve million households live in the floodplain. The USDA identifies 153 million acres of rural land that experience some flooding. We don't necessarily need to know the absolute value of these numbers at any given time, but we would like to know if and how they are changing. And they become very important when we consider managing floodplains.

Uses of floodplains are important to our management goals. We have 50 million acres of cropland, apparently, and twelve million households at risk because they're located in a floodplain. On the other hand there may be 44 million acres of natural forested floodplain where the ecologic definition of floodplains is at least partially satisfied, and another 49 million acres of pasture and range land where it easily could be. These are the floodplains. What kind of management do we do here, and why?

References

1. Walker WR, Hrezo MS, Haley CJ. Management of water resources for drought conditions. National Water Resources Summary, 1988-1989. Denver: USGS, 1991:150.

2. Hoffman WL, Campbell C, Cook KA. Sowing Disaster. The Implications of Farm Disaster Programs for Taxpayers and the Environment. Washington DC: Environmental Working Group, 1994:3.
3. Williams PB. Flood control vs. flood management. Civil Engineering May 1994:54.
4. Schneider SH, Temkin RL. Water supply and the future climate. Climate, Climatic Change and Water Supply, Chapter 1. Washington DC: National Academy of Science, 1977.
5. National Water Summary 1988-89. Hydrologic Events and Floods and Droughts. USGS, 1991:6.
6. Stockton C. Paleoclimatic data. Climate, Climatic Change and Water Supply, Chapter 2. Washington DC: National Academy of Science, 1977.
7. LR Johnston Associates. Floodplain Management in the United States: An Assessment Report, Volume 2. Federal Interagency Floodplain Management Task Force, 1992:6-3.
8. *Ibid.*
9. National Water Summary, 1988-89, *op cit,* 5.
10. *Ibid.* Hirschboeck KK. Climate and floods, 67.
11. Handmer J, ed. Flood Hazard Management: British and International Perspectives. Norwich UK: Geo Books, 1987.
12. Johnston, *op cit,* 1-6.
13. Moore DP, Moore JW. The Army Corps of Engineers and the Evolution of Federal Floodplain Management Policy. Boulder: Institute of Behavioral Science, University of Colorado, 1989:6.
14. Johnston, *op cit,* 6-7.
15. *Ibid,* 6-8.
16. *Ibid,* 6-9,10.
17. Beven K, Carling P, ed. Floods: Hydrological, Sedimentological and Geomorphological Implications. Chichester, United Kingdom: John Wiley & Sons, 1989:3.
18. Interagency Floodplain Management Review Committee, Sharing the Challenge: Floodplain Management into the 21st Century. Report to the Administration Floodplain Management Task Force, 1994:60.
19. Johnston, *op cit,* 1-3.
20. *Ibid,* 1-5.
21. Federal Interagency Floodplain Management Task Force. A Unified National Program for Floodplain Management. Washington, DC 1994:3.
22. Johnston, *op cit,* 3-3.
23. *Ibid,* 3-2.
24. *Ibid,* 3-12.
25. *Ibid,* 1-6.

CHAPTER 3

THE NEED FOR FLOODPLAIN MANAGEMENT

WHY MANAGE AT ALL?

Floodplains have been managed, historically, to prevent damages. It is on the floodplains that overbank flooding does what is normally called "economic" damage. In the spring, particularly, when the amount of water pouring out of the upper watershed is greatest, it spreads out across the floodplain causing damage to the people, their activities and their property located there. People have always built their homes, grown their crops, worked and played on floodplains where these things can be done better than almost anywhere else. Floodplains provide flat and fertile ground that is very productive and easy for farmers to work. Businesses and industry once depended heavily upon the inexpensive access to river transportation available in river towns and many still benefit from it today. Rivers provide a source of water and a place to dump the waste. Floodplain land provides a cheap and easy location for residential development precisely because it floods. Floodplain homes and recreational facilities benefit from the beauty of streams and lakes, recreation in and on the water and access to wildlife for hunting, fishing or for simply watching. All this activity is disrupted, sometimes disastrously, by flooding. One of the most important goals of floodplain management, in fact the only one in many minds, is to prevent the disruption to activity, damages to homes and businesses, and even loss of life which flooding causes.

Another important interest in the floodplain is an environmental one. The vital role which wetlands and floodplains play in healthy ecosystems requires that they be protected where they still exist and restored where they do not. The diversity and abundance of a wide variety of vertebrate and invertebrate animals, birds and fish, along with the vegetation which supports their lives, depends upon the special conditions of the natural floodplain. We put a value on these natural systems for several reasons, only a few of which are very specific. Sportsmen require strong and healthy communities of animals, birds and fish. Food industries depend upon the nurturing capacity of floodplain conditions. More important yet less specific is the value we place on protecting the natural environment, our general anxiety about further loss of wildlife habitat and the loss we feel when species go out of existence forever. It is not possible to put a dollar value on these poorly articulated concerns, but they are real enough that we are willing to spend money on them. Protection and preservation of the precious ecosystems that flourish on the floodplain is a second and important floodplain management goal.

The need for floodplain management is spurred, therefore, by economic and ecological interests that basically are incompatible. Ecosystems thrive from overbank flooding and economic activities do not. Recreational interests fall into both camps. Activities such as hunting, camping, rafting and bird watching are served by preservation of the natural floodplain. Other water-based activities such as swimming, jet skiing and power boating are better served by large stable bodies of water with permanent and predictable shorelines. The floodplain management issue of today is how to accommodate both sets of interests.

ECONOMIC IMPACTS OF FLOODING

Flooding actually provides some economic benefits. In earlier low-technology times floodplain agriculture took advantage of the enrichment of floodplain soils from deposited nutrients. Our Mesopotamian ancestors drew from the floodwater of the Tigris and Euphrates Rivers to irrigate the great alluvial fan on which they built their ancient civilization, and trapping the floods of the Nile to grow crops in Egypt and Sudan dates back at least 6000 years. Even today many floodplain dwellers in underdeveloped countries adjust their agricultural activities to accommodate the annual floods. Recent flood control projects in Bangladesh, to the extent to which they have successfully protected the floodplain from encroaching floods, caused reductions in soil fertility, greater dependence on fertilizers and decline in fish populations.[1] Flooding provides, as well, for the expansion and protection of economically valuable spawning areas and fisheries, and the flooding process with its shallow ebb and flow across wide expanses cleanses the water of sediments and harmful pollutants. Finally, the very capture of flood waters by the plains upstream can serve to retard water and reduce flood crests further below in the watershed, preventing downstream damages. The economic benefits of floods, however, are rarely quantified and generally ignored.

The economic damages caused by floods, on the other hand, are specified with a precision that implies a far greater connection with reality than actually exists. The measurement of the economic damages caused by flooding serves two purposes. The first is to assess the real impact of an historical flood. The second is to predict the damages from a hypothetical flood, essential in the evaluation of a potential flood control project or development of rational floodplain management strategies.

Economic damages produced in the floodplain by the inundation from flooding have traditionally been limited to property damage and loss of agricultural income. Loss of life, once a significant impact of overbank flooding, is less frequently associated with riverine flooding today, although alluvial fan and flash floods as well as dam failures may still take their tolls. In the 1993 floods in the upper Mississippi River basin, for example, 47 people are claimed to have lost their lives. Of the 33 whose cause of death is documented, 19 got trapped by floods while traveling in cars, some knowingly taking risks and others surprised by localized flash flooding events. Four boys and two adults were lost while exploring caves in a park that had been officially closed, one man was electrocuted, another lost in a boating accident and five drowned when on foot near floodwater. In one case a house was actually swept away by floodwater, killing one older woman. Only this last case and perhaps several of the car-related deaths were those of totally helpless victims. The others had, to varying degrees, knowingly put themselves in danger. Today people are alerted early enough to get out before the floods occur, although they rarely can take their property with them. We have come a long way since 1927, when Mississippi River floods destroyed 41,000 buildings and killed between 250 and 500 floodplain residents, and we are a long way from Bangladesh where 3,000 died in floods in 1988.

The reported property damages have been escalating since the turn of the century (Fig. 3.1), yet estimates of damages caused by large floods are remarkably inaccurate. This is inevitable, since no agency or office has been charged with responsibility for damage assessment, or given the funding to do it. The Federal Emergency Management Agency (FEMA) does only enough of an assessment, immediately after a natural disaster strikes, to determine that the damages are sufficient to warrant a presidential disaster declaration and thus be eligible for federal aid. The official estimates come from the National Weather Service (NWS) which puts together, as best it can, a back-of-the-envelope tabulation from a variety of state and local sources, including media reports. The NWS has no

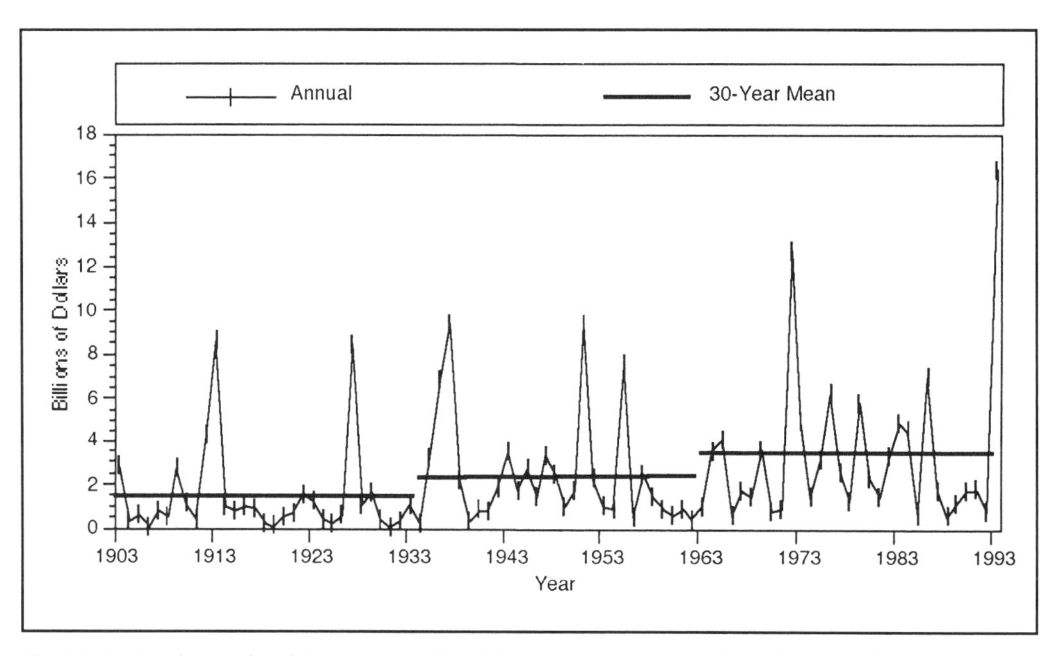

Fig. 3.1. National annual and 30-year mean flood damages since 1903, adjusted to 1993 dollars. The mean annual damage increased from $1.5 billion in the first 30 years to $3.4 billion in the most recent 30 years. (Source: U.S. Weather Bureau). Reprinted with permission from the report "Revisiting Flood Control: An Examination of Federal Flood Control Policy in Light of the 1993 Flood Event on the Upper Mississippi River," by Wetlands Research, Inc., Chicago, Illinois.

formal responsibility and certainly not the resources for this function and this agency is the first to recognize its gross inadequacy. This unsystematic collection of damage data produces some rather bizarre results. The 1993 flood on the upper Mississippi River is a case in point.

Estimates of damages from that meteorological event ranged, during the twelve months following the floods, from $10 to $20 billion. The official number finally became $15.7 billion, the NWS's best guess. These damages were much larger than those from any prior event. The average annual flood damages between 1951 and 1985, for example, were $2.18 billion.[2] A flurry of federal activity followed the 1993 disaster, including a directive from the Federal Interagency Task Force on Floodplain Management to a special Floodplain Management Review Committee chaired by General Gerald A. Galloway to produce a report. The "Galloway Report,"[3] published in June 1994, gave the stamp of approval to the $15.7 billion figure although it also lamented the general lack of reliable damage estimates. The Galloway Committee produced a comprehensive analysis of floodplain management policy and recommendations for the future which apparently prompted the issuance, in 1994, of the most recent version of a Unified National Policy for Floodplain Management.[4] It seemed clear that floodplain management policies were considered to be connected in important ways to the $15.7 billion of damage, the implication being that if better policies had been in place these damages could have been at least partially avoided. But that is not quite the case (see Fig. 3.2).

The 1993 floods on the upper Mississippi River also resulted in a Floodplain Management Assessment (FPMA)[5] prepared by the Chicago District Corps of Engineers (Corps) in response to a congressional call for action after the floods. This study was to include the publication of a separate Economic Damage Data Collection Report in February 1995[6] generated by the Corps' Lower Mississippi Valley (LMV) Division. The main report came out in mid-1995 but in early 1996 the damage data report was still unavailable and no publication date had been set. The raw data that was available in draft form and the aggregate numbers used in the FPMA report, however, demonstrate substantial discrepancies from the official numbers.

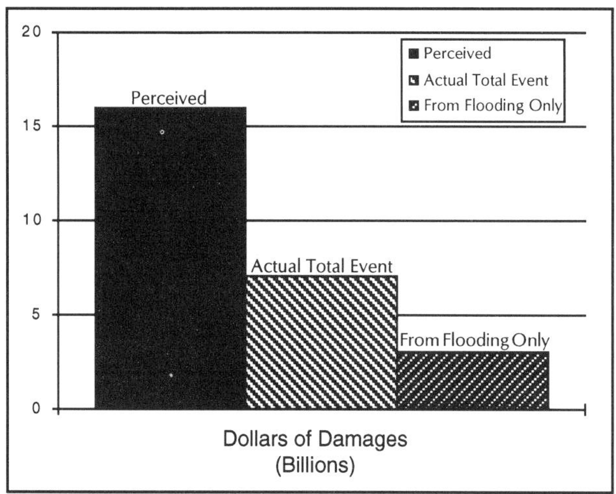

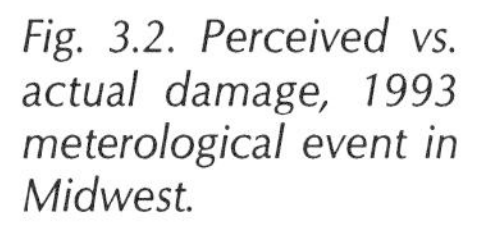
Fig. 3.2. Perceived vs. actual damage, 1993 meterological event in Midwest.

The main FPMA report recognized that an analysis of floodplain management issues had to distinguish between two associated but different damage-creating phenomena in 1993: the overbank flooding along the rivers and the saturation of soils in the upper watershed. Five-hundred and five counties were declared eligible for disaster assistance, that summer, but the FPMA found that only 125 of them were adjacent to the major rivers in the devastated area and therefore had been subject to overbank flooding. They quite appropriately restricted their assessment of floodplain management to them. They identified \$3.09 billion of damages from actual overbank flooding: \$978 million, or 32%, of agricultural and other rural losses, \$662 million or 21% of residential damages in urban areas and 47% or \$1.447 billion of "other urban damages" (Fig. 3.3). No further detail is provided for the set of 125 counties. The draft damage data available from the LMV, although it breaks out damages by categories for each of the 450 counties in the Mississippi and Missouri River basins, does not observe the important distinction made in the main report. Within the "other urban" category that is so large for overbank flooding, the breakdown for the damages in the entire 450-county population is 46% commercial and industrial; 27% transportation; 19% public and 8% utilities (Fig. 3.4). Whether this subdivision of "other urban damages" would apply to the 125 counties on the floodplain is anybody's guess. The FPMA is discussed in more detail in chapter 9.

The distinction between the entire hydrometerological event in 1993 and the overflows from the main river channels was not lost in most informed discussions about the flooding, even though its implications were. The Galloway Report went so far as to report that "much" of the $15.7 billion of damages was in the upper watershed, the result of soil saturation, surface ponding and high groundwater levels. Considering the extent of upland flooding, the report suggested that over half the total damages were agricultural.[7] The six volume Corps Post-Flood Report[8] skirted the issue, quoting the NWS figure (which had now become $15.6 billion), giving some anecdotal damage information and stating that the June 1995 FPMA would provide the damage data.[9]

The distinction between flood and upland damages, however, accounts for only part of the problem. The draft damage figures generated by the LMV for the entire event total only $6.5 billion,

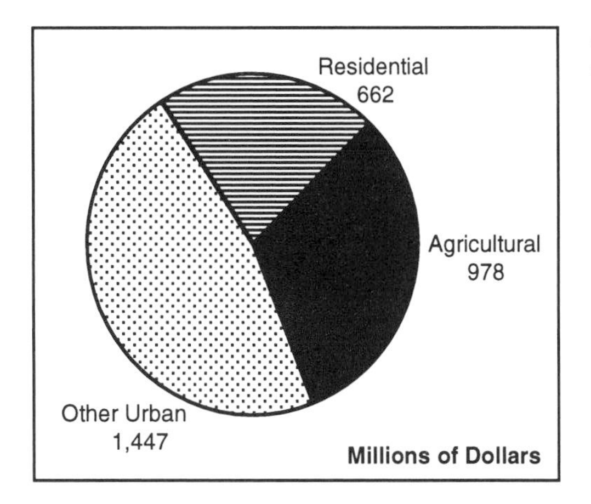

Fig. 3.3. Types of damage, 1993 overbank flooding.

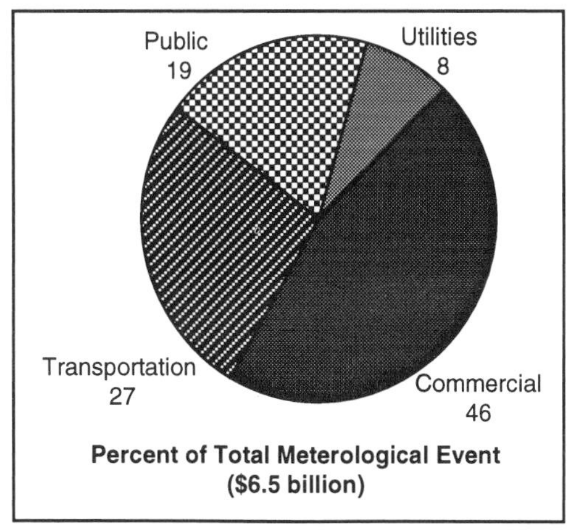

Fig. 3.4. Types of urban other damage, 1993 Midwest floods.

less than half the $15.7 figure that has gone into the official record book. The distribution of this $6.5 billion among the damage categories is roughly 60% agricultural, 12% residential and 26% other urban damages. Even the little data that is available shows other discrepancies with prior reports. The Galloway Report, for example, estimates that 50,000 homes were damaged from riverine flooding and another 50,000 from groundwater or sewer backup.[10] The LMV draft data indicate that a total of 36,781 residences and 3,791 commercial buildings experienced damage from the entire hydrometeorological event.

The widespread perception of the impact of the long lasting and intense Midwestern rainstorms of the summer of 1993 is that they caused almost sixteen billion dollars of flood damages of the kind that are amenable to floodplain management and that the majority of those were agricultural. The truth lies closer to a maximum three billion dollars of damages on the Midwestern floodplains, of which over three quarters was urban. The inaccurate reporting of flood damages is not surprising to anyone working in the field. Howe and Cochrane quote Thomas P. Grazulis as saying "I haven't found anyone within the flood research community [who] actually believes the NWS numbers" in a memo to the National Science Foundation (NSF) Advisory Committee on February 2, 1989.[11] What is surprising is that the problem has not been corrected. Such a gross misrepresentation of a problem is unlikely to lead to appropriate, much less efficient, solutions.

Howe and Cochrane go on to propose a framework for categorizing and collecting flood damage data. They suggest that monetary measurements be made of economic damages to property and people and of interruptions in production processes caused by floods. (Two other of their damage categories, historical monument and natural capital, are discussed below.) Under their system property damages would include business and government capital assets and goods, residences and personal property. They propose measuring damages to people not only in terms of loss of life but also of injuries that cause loss of productivity in the household and production activities as well as the "disutility of physical and psychological malaise."[12] Interruptions in production would be calculated as delays or total loss of "value added" in activities such as agricultural production, commercial fishing, manufacturing, transportation systems, service industries, government operations and services. Howe and Cochrane propose using the 93-category U.S. Standard Industrial Classification to identify economic activities, with the addition of a household sector that would include such damages as interrupted food preparation and relaxation.

The LMV draft data was summarized in damage categories of residential, commercial/industrial, public, transportation, utilities, agricultural, emergency and other. These in turn were derived by aggregating a set of approximately 50 data units that include numbers of units damaged, revenues lost and property destroyed for a larger number of categories than is generally examined. Usually damage reports, if they are categorized at all, are divided simply into urban damages, more or less a catch-all for destruction of property, and agricultural damages, the loss of that season's income from crops. More frequently they appear as a single lump sum.

It's entirely appropriate to represent a "best guess" for damages as a single figure, but true damage estimates require division and definition. It is not appropriate, however, to lump "excess moisture" with overbank flooding damages, because the tools we use to solve the two problems are as different as are the problems themselves. Historical flood damage estimates (see Fig. 3.1) may be much closer to the mark for entire hydrometerological events but relate neither in magnitude nor trend to overbank flood damages. Hoffman has reported, for example, that for every $1 in crop insurance payments for flood damages, $8 are distributed to compensate for "excess moisture" problems.[13] While an understanding of where and how the damages from floods actually occur is the first essential step in mitigating them, it is a step that we have not yet taken.

ECOLOGIC IMPACTS OF FLOODING

Flooding has a positive and valuable impact on the natural systems which it affects, although extreme flood events may cause limited ecologic damage. Unusually high water levels of long duration have been observed to damage individual stands of vegetation. Large numbers of sycamores, for example, were destroyed by the 1993 floods in the upper Mississippi River valley, as were all the hackberry and sugarberry at two navigation pools surveyed by the Corps.[14] Extreme flood events will always cause some short-term ecologic setbacks; some habitats will be destroyed, even as others are being created and the individuals of certain plant and animal species may perish in events that benefit their species as a whole. Harmful ecologic impacts from flooding are, however, as much of an anomaly as are the beneficial economic ones.

From an ecologic point of view the impacts of flooding are overwhelmingly positive. The 1993 flood is described as "a boon to many kinds of plants and animals."[15] River systems are dynamic systems, characterized by process and change from every point of view: geomorphologic, hydrologic and ecologic. As water moves down through the watershed, soils are shifted, rock is scoured, banks are eroded, sediments are deposited, new channels are formed and habitats are created and destroyed. Every season, in fact every day, sees variation in the water cycle: new patterns of precipitation, evaporation, transpiration, seepage and flow evolve. It is these dynamic processes, far more than static observations of a river system such as channel cross sections or floodplain profiles, that characterize the system.

The fluid nature of a watershed energizes its ecologic systems. The value of flooding is best described in terms of the *flood pulse,* the lateral exchange between the river channel and its floodplain made possible by annual flooding, an exchange that itself is unique each year as flood flows and boundaries of inundation vary. The NRC report, *Aquatic Ecosystems,*[16] argues that "rivers and their floodplains...are so intimately linked that they should be understood, managed and restored as integral parts of a single ecosystem."[17] The report proposes that floodplain management for ecologic purposes should be organized to protect and enhance the following critical features of a healthy river system:[18]

- Flow. The flow transports nutrients, sediments, pollutants and organisms in one direction, downstream, permitting constant mixing.
- Openness. No boundaries inhibit the exchange of materials and energy.
- Dynamism. There is constant change and disturbance within a longer term dynamic equilibrium.
- Patchiness. Change produces longitudinal patchiness in riffles and pools; lateral patchiness in backwaters and

eddies; and vertical patchiness from above to below the riverbed.
- Resistance and resilience. The resistance produced by constantly moving water builds up resilience and strengthens the biota living in that environment.

Thus the dynamics and diversity of a river system promote the dynamics and diversity of its natural ecosystems. It is the very mix of riffles, pools, sandbars, oxbows, islands, side channels, sloughs, backwaters, etc., all with different depths, velocities and substrates, continually forming and reforming the channel configuration and the floodplain profile, that stimulate the growth and reproduction of all manner of living organisms living there.

Although the NRC report points out important differences between the energy transfer processes of the river-floodplain systems of major water courses and those of riverine-riparian systems of smaller watersheds,[19] an important stimulus is provided by the conditions of shallow water and slow flow. Shallow depths allow the penetration of light to activate photosynthesis, the use of sunlight to transform inorganic substances into plant material, and produce oxygen as a byproduct. Slow flows may reduce erosion, allow the sediments, heavy metals and pollutants to settle out, permit the gradual release and exchange of nutrients between the channel and the floodplain and facilitate the recharge of groundwater aquifers. Wetland vegetation, an efficient converter of solar energy,[20] reduces shoreline erosion, stabilizes river banks, and processes chemical and organic wastes.

The most important function of the floodplain is to provide transition between terrestrial and aquatic system ecosystems. Viewed in the context of the life-and-death-cycle,[21] nutrients are released from the floodplain and exchanged with the river; aquatic plants and invertebrates grow; plant matter decays on the floodplain and is exported to the river; plankton produced in floodplain depressions, along with bottom organisms and aquatic plants, go into the river; fishes travel from the river to the floodplain where they spawn, grow and move back to the river; those that are stranded provide food for birds and animals (see Fig. 3.5).

No one has devised, to date, a successful system for measuring the ecological impacts of flooding. The very nature of the benefits defies the system of quantification used to measure economic impacts. Economic damages can be measured against market values, but there is no marketplace for ecosystems or their components. Two problems arise: the difficulty of quantifying the most important ecosystem benefits and, even in those instances where measurements can be taken, the impossibility of assigning monetary values to them.

To quantify the ecological benefits of flooding, measurements can be taken, for example, of numbers of species and size of populations. If the baseline data is accumulated then changes caused by a given flood can be calculated for these parameters. With enough data, models could be devised to represent particular floodplain environments and their response to different floods but good science might require an infinite number of variations. Each floodplain environment has its unique characteristics: the broad low plains of the upper Mississippi River valley and the high energy tumbling American River, for example, have very little in common in natural settings and ecological values, much less flooding characteristics.

It is difficult to imagine how we could measure the truly important impacts of flooding on the ecosystem, the long-term or far reaching ones: how yesterday's flood affects tomorrow's ecosystem or how overbank flows in one small watershed might ultimately affect the entire basin. Howe and Cochrane[22] have suggested that a complete system of hazard impact assessment should include a category of natural capital where changes in ecosystem structure and function and other conditions could be measured against baseline data for a set of categories such as the following:

- Species composition, abundance and growth form and stature
- Feeding relationships among species
- Ecological dominance and key species
- Species diversity
- Indicator species and ecological indicators
- Size, shape and heterogeneity of disturbed areas
- Physical factors
- Productivity
- Trophic structure and energy flow
- Nutrient relationships

- Decomposition processes
- Succession and development of communities
- Individual species characteristics
- Disturbance event characteristics
- Evidence of other or potential natural hazard events
- Evidence of or potential for cumulative impacts
- Evidence of patterns and rates of recovery

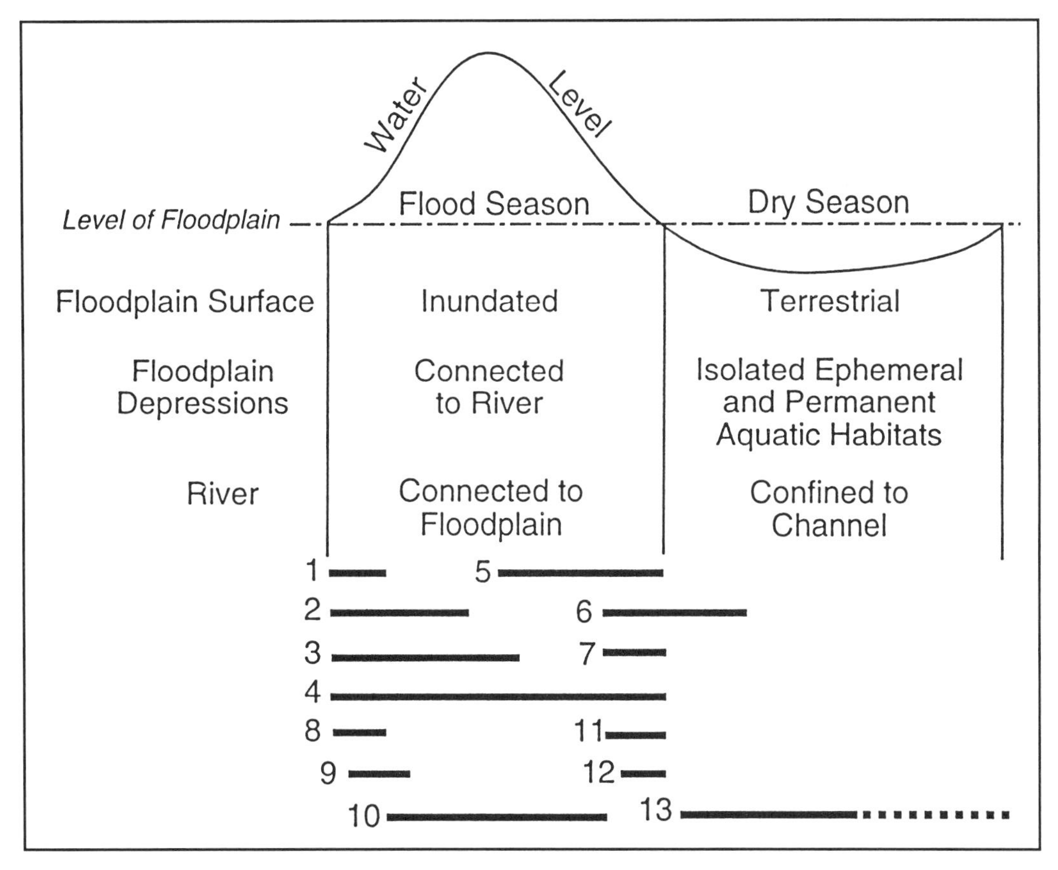

Fig. 3.5. Idealized changes in water level over an annual cycle for a large floodplain river. Numbered horizontal bars indicate characteristic annual periodicity patterns for some major interactions as follows: (1) nutrients released as floodplain surface is flooded; (2) nutrient subsidy from connected river; (3) rapid growth of aquatic plants and invertebrates on floodplain; (4) major period of dead plant matter decomposition on floodplain; (5) organic matter exported to the river; (6) maximum plankton production in floodplain depressions; (7) drift of plankton, bottom organisms and aquatic plants to river; (8) fishes enter floodplain from river; (9) major period of fish spawning on floodplain; (10) period of maximum fish growth; (11) fishes move from floodplain to river; (12) heavy fish predation losses to other vertebrates at mouth of drainage channels; (13) high mortality of fishes stranded in floodplain depressions provides forage to birds and mammals. Reproduced from Science for Floodplain Management into the 21st Century, Preliminary Report of the Scientific Assessment and Strategy Team. Report of the Interagency Floodplain Management Review Committee to the Administration Floodplain Management Task Force.

- Characteristics of neighboring ecosystems and organisms
- Evidence of controlling or limiting environmental factors

If this list correctly represents the measurements that should be made to properly evaluate the ecologic effects of flooding, it bears little resemblance to what is being done today. Although most flood control and floodplain management reports and studies contain the obligatory section on ecology, it is purely perfunctory. The FPMA prepared in 1995 devoted a separate appendix to an Environmental Resources Inventory which purported to inventory the natural resources of the region and to describe the effects of the 1993 flooding on them.[23] Almost all of the 400 pages were devoted to the inventory; only four pages addressed the effects of the flooding, less than half of which provided actual information, albeit anecdotal, on the effects of the floods on the environmental resources. Another National Research Council's Report on Flood Risk Management and the American River basin[24] points out that while environmental issues are at the very heart of the enormous controversy surrounding the recent reports on flood control alternatives in the basin, those reports have been deficient in clearly laying out the environmental impacts related to the proposed actions.[25] The report concluded, in its broader findings, that "Traditional environmental impact statements fail to evaluate flood risk management alternatives in an ecosystem context because they use a species-oriented framework. This approach has limited usefulness."[26]

It could be argued that an engineering organization such as the Corps is not particularly well equipped to perform ecologic assessments. It also could and has been argued that the Corps is reluctant to do so. It is more than likely that engineers and economists accustomed to dealing with numbers and well-tested methodologies feel awkward handling the spotty data and tentative procedures that are commonly used to assess ecologic impacts.

If measuring these ecologic impacts is difficult, monetizing them is impossible. Consider some examples of such measurements:

- Fish yield per unit water area as increased by predictable flood pulse, called the "flood pulse advantage"[27]
- Percentage of U.S. commercial fisheries landings consisted of fish and shellfish that depend on wetlands;[28]

- Counts of threatened and endangered species;
- Presence or absence of migratory corridors and stop-overs for feeding;
- Counts of floral and faunal species;
- Percentages of species of fish, crayfish and mussels that are rare, imperiled, or actually extinct;[29] and
- Recreation days for trapping, hunting, fishing, picnicking, jogging, bird watching, camping, hiking, canoeing, kayaking, jet skiing, power boating and sailing.

There has been no agreement on how absolute dollar values can be assigned to data sets such as these. Surveys are conducted, sometimes, asking how far a person would travel or how much he would be willing to pay for such commodities. Recreational and sport hunting days have been counted and valued in dollars. Yet who makes the value judgments and how? Is the last extant species of crayfish given a higher value than the tenth from last and by how much? Is a recreational day of white water rafting more or less valuable than a day of jet skiing on a reservoir? To whom and why? If biodiversity has a value, and most agree that it does, no one can say exactly what that value is. And in what geographical context is a value assigned: local, regional, national or even global? On the American River, for example, the riparian habitat assumes a greater value in the context of the state of California where less than 5% of the original riparian habitat remains. And if the river rapids hypothetically had no value to the local residents, how could we factor in the international perspective on Class IV and V rapids?

Researchers have tackled and will continue to grapple with the problem of what price tag to put on environmental phenomena. Heimlich and Langer assigned a dollar per acre value to certain wetland functions: $6,225 per acre for a waste assimilation function in a Virginia tidal marsh; $38 per acre for fish, wildlife and recreation in the Charles River and $1108 per acre for water quality enhancement in the Alcovy River, Georgia.[30] Values such as these probably have limited use. Assigning values to make comparisons is much more useful. Ranking environmental features along a natural values continuum such as the ones devised by the USDA for their Wetland Reserve Program provides a reasonable basis for selecting one property over another.

King and others tackled the problem recently[31] in the context of sustainable watershed management. They provide an interesting analysis of natural resources accounting and make a useful distinction between high and low order natural resources. High order are directly exploited by people and include such items as timber and fish. Low order resources play more fundamental roles in nurturing ecologic processes.[32] The authors caution, however, that "for resources accounting purposes the valuation problem is insoluble," and although they hold open hope for efforts to assign economic values to high order resources, they do not expect that it could be done for the low order ones.[33]

The National Performance Review conducted by Vice President Al Gore, in the early days of the Clinton administration, included a review of Environmental Management Programs.[34] The first and perhaps the most important of the four environmental recommendations made in this review was to improve federal decision making through environmental cost accounting. The Vice President's report proposes that the U.S. Environmental Protection Agency and the Corps initiate environmental cost accounting demonstration projects and that, subsequently, environmental cost accounting guidelines be developed for implementation, through a presidential directive, throughout the federal government.

The search will continue for a methodology that will translate environment values into monetary ones as long as floodplain management decisions are driven by benefit/cost ratios. And, as we shall see in following chapters, the law requires it. Economic values drive the process. Until a system is devised that methodically records and quantifies the impacts of flooding on ecosystems that has anywhere near the precision used to evaluate economic damages, the ecologic side of the equation will suffer.

OTHER IMPACTS

Certain recreational interests may be negatively impacted by floods to the extent that recreational facilities are destroyed or rendered temporarily useless. Other types of recreation are dependent upon natural environments and/or thriving communities of animals, fish and plants and these are benefited by the ecologic nurturing qualities of the flood pulse. The constituencies for these recreational activities may play important roles in decision making, as on the American River where upstream white water rafters threw their weight behind the effort that successfully stopped the Corps' 1991 proposal (described in more detail in chapter 9).

In summary, floods cause economic damages and provide ecologic benefits. We manage floodplains to reduce the former and to protect the latter. The ways in which floods enhance ecologic systems are complex and far reaching, and we have not yet found ways to successfully describe and measure them. The economic damages that floods produce are easy to describe and measure, but the data is difficult and expensive to collect which is, perhaps, the reason we have been unwilling to do that either. There is agreement, however, that floodplains must be managed to meet these goals and that it is in the public interest to do so. It hasn't always been seen in quite that way.

REFERENCES

1. Williams PB. Flood control vs. flood management. Civil Engineering May 1994:54.
2. LR Johnston Associates. Floodplain Management in the United States: An Assessment Report, Volume 2. Federal Interagency Floodplain Management Task Force, 1992:3-20.
3. Interagency Floodplain Management Review Committee. Sharing the Challenge: Floodplain Management into the 21st Century. Report to the Administration Floodplain Management Task Force, 1994.
4. Federal Interagency Floodplain Management Task Force. A Unified National Program for Floodplain Management, Washington, DC 1994.
5. United States Army Corps of Engineers. Floodplain Management Assessment of the Upper Mississippi River and Lower Missouri Rivers and Tributaries. 1995.
6. *Ibid*, 1-4.
7. Interagency Floodplain Management Review Committee, *op cit*, 15.
8. United States Army Corps of Engineers. The Great Flood of 1993 Post-Flood Report: Upper Mississippi River and Lower Missouri River Basins. Chicago: North Central Division, Corps of Engineers, 1994.
9. *Ibid*, 40.
10. Interagency Floodplain Management Review Committee, *op cit*, ix.
11. Howe CW, Cochrane HC. Guidelines for the Uniform Definition,

Identification and Measurement of Economic Damages from Natural Hazard Events. Boulder: Institute of Behavioral Science, University of Colorado, 1993:1.
12. *Ibid,* 12.
13. Hoffman WL, Campbell C, Cook KA. Sowing Disaster. The Implications of Farm Disaster Programs for Taxpayers and the Environment. Washington DC: Environmental Working Group, 1994:2.
14. Interagency Floodplain Management Review Committee, *op cit,* 31.
15. Sparks R, Sparks R. 1994, Ecosystem restoration and the flood of 1993. USA Today Magazine, July 1994, 40-42.
16. National Research Council. Restoration of Aquatic Ecosystems. Washington, DC: National Academy Press, 1992.
17. *Ibid,* 175.
18. *Ibid,* 178.
19. *Ibid,* 185.
20. LR Johnston Associates. Floodplain Management in the United States: An Assessment Report, Volume 2. Federal Interagency Floodplain Management Task Force, 1992:2-9.
21. Interagency Floodplain Management Review Committee, Science for Floodplain Management into the 21st Century. Preliminary Report of the Scientific Assessment and Strategy Team to the Administration Floodplain Management Task Force, 1994: Ch 6.
22. Howe CW, Cochrane HC. *Ibid,* 15-16.
23. U.S. Army Corps of Engineers, 1995, Appendix C.
24. National Research Council. Flood Risk Management and the American River Basin; An Evaluation, Washington DC: National Academy Press, 1995.
25. *Ibid,* 7.
26. *Ibid,* 210.
27. Interagency Floodplain Management Review Committee, Science for Floodplain Management into the 21st Century. Preliminary Report of the Scientific Assessment and Strategy Team to the Administration Floodplain Management Task Force, 1994:128.
28. Johnston, *op cit,* 2-17.
29. National Research Council. Restoration of Aquatic Ecosystems. Washington, DC: National Academy Press, 1992:177.
30. Johnston, *op cit,* 2-3.
31. King DM, Bohlen CC, Crosson PR. Natural Resource Accounting and Sustainable Watershed Management. Solomons, MD: University of Maryland, 1995.
32. *Ibid,* 11.
33. *Ibid,* 22.
34. National Performance Review, Reinventing Environmental Management. Washington, DC: Office of the Vice President, 1993.

CHAPTER 4

Defining the Public Interest

WHO MANAGES FLOODPLAINS?

Today the federal government is expected to play the dominant role in floodplain management, but that has not always been the case. The management of floodplains in the United States began as a private initiative, with farmers building their own levees and river towns raising flood walls. The states jealously guarded their sovereign rights and river projects were considered to be the purview of the federal government only to the extent that interstate commerce was dependent upon open navigation channels. Although flood protection was provided sporadically, under the guise of navigational aids, it was not until the 20th century that Congress officially approved flood control projects and acknowledged, in 1936, that it was in the national public interest to prevent economic damages on the floodplains of all our rivers.

Even within its short history, floodplain management has been transformed in several ways. What began as *flood control,* controlling floods by building structures to contain them, has broadened to include *floodplain management,* controlling and restricting the uses of our floodplains. Structural flood control projects such as dams and levees have been enhanced by nonstructural measures which reduce or prevent flood damages without containing the flood waters themselves. The relatively recent attention to the natural environment has spawned an awareness that flood control can cause environmental damage. The importance of riverine wetlands has been recognized and regulations have been adopted to protect them. Floodplains are managed today to accommodate the two sets of public interests, economic and ecologic.

ESTABLISHING THE FEDERAL INTEREST

At the beginning of the 19th century, flood control was not acknowledged to be a proper activity of the federal government. On the contrary, the federal flood control functions we perform today were then perceived to be in violation of the federal constitution. Projects on the rivers, including those which aided navigation, were classified as "internal improvements," an activity retained by the states. The states, however, had become too poor to take on additional duties, having lost their former revenues from customs collections to the federal government.[1] It was only after the 1824 Supreme Court decision, *Gibbons v. Ogden*, when the federal government's power under the Commerce Clause "*To regulate Commerce... among the several States*" was expanded to include the financing and construction of road and river improvements, that such projects began to be built with federal funds. Although the battle had not yet been won for flood control projects, Congress was under pressure each time it flooded and gradually began to give legislative approval to navigational projects whose clandestine intent was flood control. In 1849 and 1850, the passage of the Swamp Land Acts followed close upon flooding on the lower Mississippi River. These Acts authorized land grants, transferring more than 100,000 square miles of floodplain from the federal to state governments with the understanding that the revenue from their sale was to be used to build levees and drainage canals. An 1874 flood led to the creation, in 1879, of the Mississippi River Commission, which initiated the program of levee construction (explicitly denying a flood control function in each piece of enabling legislation) which had been recommended in the 1871 *Report on the Physics and Hydraulics of the Mississippi River* by Captain Andrew A. Humphreys and Henry L. Abbot and which was to dominate flood control thinking into the next century.

Meanwhile, contiguous agricultural property owners in flood prone areas were forming drainage districts, legal entities which taxed themselves to build, maintain and operate drainage and levee systems to protect their crops from water damages. Between 1859 and 1959 these specialized local government units drained more than 117 million acres of wetlands, in the uplands as well as the floodplains. The most intensive period of activity was between 1900 and 1920, when over half these private lands were drained.[2]

After the turn of the century, a series of flooding disasters (see chapter 2) prompted some precedent-setting congressional actions. Populated areas had been hit in the 1907 Pittsburgh and the 1913 Ohio River floods, and in 1912 the 250 million cubic yards of earthen levee that were considered invincible protection for the lands along the Mississippi were breached.[3] The House Committee on Flood Control was created in 1916. Immediately thereafter, in 1917, Congress passed the first legislation explicitly for flood control, restricting it to the control of floods on the lower Mississippi and Sacramento (California) Rivers and requiring local interests to pay one-third of all the costs. In 1927, the U.S. Army Corps of Engineers (Corps) was authorized to do what was known as the "308" river basin planning studies, providing that agency with the opportunity and funding to collect hydrologic data which would position it, in 1936, to take charge as the country's lead flood control agency. At the direction of President Herbert Hoover, the boundaries of Corps districts had already been redrawn in closer alignment with river basin boundaries.[4] In 1928, in response to the previous year's latest Mississippi floods, the Mississippi River and Tributaries legislation was enacted that reflected some changes in congressional thinking. For the first time floodways, spillways and channel improvements were authorized in addition to levees, and local interests were relieved of all financial responsibility. The rationale for increasing the federal financial share was more the inability of the local interests to shoulder the costs than it was the clear conviction that broad public benefits would accrue.

Thus the stage was set for passage of our country's first comprehensive flood control act. Precedents had eased the way toward accepting the constitutionality of federal flood control, the necessity of federal funding had been acknowledged, the technology had been expanded to include a variety of structural programs and the Corps was in a position to take the bureaucratic lead. All that was needed was the impetus and the floods of the early 1930s provided it, helped along by the economic troubles of the country.

THE FLOOD CONTROL ACT OF 1936

During the early 1930s the country was staggering under the effects of the Great Depression. Presidents Herbert Hoover and Franklin Delano Roosevelt both saw flood control works as employment relief measures, and Work Projects Administration (WPA) workers were used on Corps projects, which were probably valued more for the jobs they generated than the waters they kept at bay. Interest in a comprehensive flood control act was building in Congress, an act which could lay the groundwork for a broad-based national flood control effort. Conflicting interests arose. Roosevelt, who had a personal interest in forestry, an appreciation of the importance of comprehensive watershed management and an aversion to pork-barrel funding, favored a nonpolitical permanent commission for water resource planning. He had created the Tennessee Valley Authority in 1933 to provide a multipurpose water resources planning approach to the Tennessee River Basin. In 1934 he created the National Resources Board (NRB), asking it to draw up a comprehensive plan for water resources management. Congress was on its own track, however, and resisted the influence of both the president and the NRB, which envisioned its own Water Resources Committee (the old Mississippi Valley Committee) as playing the lead role in future flood control programs. The legislators preferred to keep the planning and project selection for flood control, as they had navigation for many years, in the hands of the Corps. They were influenced as much by the political alliance of the NRB with the president, no doubt, as by a preference for a single-project approach to watershed management.[5] More severe flooding in 1935 stimulated Congress to move ahead. One hundred flood control bills were introduced, among them HR 8455 which, although it failed passage in that year and was amended drastically, became the basis for the Flood Control Act of 1936, signed into law by President Roosevelt on June 22.

The 1936 Flood Control Act opened with a policy statement that irrevocably established the constitutionality of the federal interest in flood control, echoing the General Welfare clause:

"Floods... constitute a menace to national welfare... flood control is a proper activity of the Federal government... in the interest of the general welfare." The scope of that interest was defined as the navigable waters of the United States or their tributaries and the authorized activities were "investigations and improvements." The bill laid down the condition that such activities would be engaged in only "if the benefits to whomsoever they may accrue are in excess of the estimated costs" and it spelled out what is known as the ABCs of local involvement: that local interests would provide the land, the easements and the rights of way, would hold the federal government harmless and would maintain the completed structures. It authorized the construction of more than 200 specific flood control projects. Finally, as a sop to the President's and the NRB's desire for broad-based water resources management, it was amended at the last minute to authorize the newly created Soil Conservation Service (SCS) in the Department of Agriculture to do flood control planning in upper watersheds.[6] The actual language of Section 2 was "That hereafter, Federal investigations and improvements of rivers and other waterways for flood control and allied purposes should be under the jurisdiction of and shall be prosecuted by the War Department...and Federal investigations of watersheds and measures for runoff and waterflow retardation and soil erosion prevention on watersheds shall be under the jurisdiction of and shall be prosecuted by the Department of Agriculture..."

EMERGING DIFFICULTIES

The Department of Agriculture never exercised the precise authority given to the SCS in the 1936 legislation, but it was eventually was given a role in flood control by the Flood Control Act of 1944 and the 1954 Watershed Protection Act. In 1954 the SCS was authorized to cooperate with local entities in "566 projects," named for the subsequent public law, to construct small dams for watershed protection and flood prevention in watersheds of 250,000 or fewer acres. Although the investments in SCS projects have been minor relative to the Corps flood control activities, this somewhat arbitrary distinction between the upper watershed and the river channels has been awkward, and the small watershed projects have represented an alternative to some aspects of Corps flood control philosophy: small dams as opposed to large ones and upstream as opposed to downstream interests. The SCS has often been used as a foil against the Corps, which in its turn has criticized SCS methods of calculating project benefits. The SCS has included the future and potential damages and therefore generated benefits for its projects that far exceed the calculations of the Corps. Perhaps the most contentious deviation from Corps policy has been that SCS projects have been built entirely at federal expense. These and other turf battles, inevitably, have found their separate constituencies, not only in the general population but also in the Public Works and Agriculture Committees in Congress.

The comprehensive long-range planning approach visualized by President Roosevelt never came to pass. The NRB was ignored by Congress in preparation of the 1936 Act and by the President when it recommended against signing the legislation. Finally in 1943 the NRB, by then renamed the National Resources Committee, was abolished. Comprehensive water resources planning was neglected by the federal government for several more decades, until the Water Resources Council (WRC) was authorized by the Water Resources Planning Act of 1965, only to be dismantled by President Ronald Reagan in 1982. The 1965 legislation also authorized the creation of River Basin Commissions, but the six that were eventually established were themselves abolished in 1981. During the half century following the 1936 legislation, flood control policy was driven by more than 100 water resource laws, several dozen executive orders, more than 50 interagency agreements and more than 60 Office of Management and Budget (OMB) circulars.[7] Twenty-three billion dollars have been spent on flood control projects,[8] more than 400 reservoirs have been built and 10,500 miles of levees have been constructed[9] by the Corps alone. Flood Control Acts were passed in 1938, 1944, 1946, 1948, 1950, 1954, 1955, 1958 and 1960. Over $150 billion in flood damages have been prevented in the first 50 years of construction after 1936, $100 billion of it on the lower Mississippi River.[10]

Yet in the years following the 1936 legislation it gradually came to be noticed that the more public money we invested in flood control, the more the economic damages were increasing. Some observers thought they knew the reasons why. In 1937 General Max C. Tyler, Assistant Chief of the Corps, blamed our flood problems on floodplain development;[11] in 1951 Arthur Maass criticized the Corps for catering to the demands of special interests and of Congress and for the absence of sound water resources planning, in *Muddy Waters, the Army Engineers and the Nation's Rivers.* A 1955 publication pointed out that in 1949, a comparatively flood-free year, the damages were greater than those from the 1903 floods which were "some of the most extraordinary... on record" *(Floods,* by William G. Hoyt and Walter B. Langbein). A 1960 report prepared by retired Brigadier General Miles M. Dawson for the State of Ohio predicted that continued development of the floodplain would create flood disasters which no amount of structural protection could prevent.[12]

At the University of Chicago's Department of Geography first Harlan H. Barrows, in the 1920s, and later Gilbert F. White, both of whom played a role in Roosevelt's New Deal, became active critics of the nation's flood control policies. White's 1945 doctoral thesis, *Human Adjustment to Floods,* republished in 1953, argued that to be successful, flood control programs had to take human encroachment on the floodplain into account, include nonstructural solutions and fully evaluate the benefits and the costs.[13] Later, in a 1958 study, White pursued the same theme and attempted to evaluate the changes in floodplain occupancy in 17 communities.[14] This study estimated that floodplain development was expanding by 2.7% annually and that "the Corps of Engineers is, against its inclinations, one of the major real estate development agencies in the country.... There is no denying the fact that the protection works do accelerate certain land use changes and make others possible."

During the 1960s, stimulated by the work of White and many others, the federal government took a number of nonstructural initiatives, including professional orientations other than engineering in the planning process. In 1965 an Interagency Task Force chaired by White made a set of important recommendations, incorporated into House Document 465, which became the first

Unified National Program for Managing Flood Losses. This document resulted immediately in an Executive Order that directed federal agencies to evaluate flood hazard potential before locating new buildings on the floodplain. It led eventually to the National Flood Insurance Program (NFIP) of 1968, and it insured that nonstructural measures would play a major role in future United States floodplain management policy. A Federal Interagency Floodplain Management Task Force was established in 1975 to coordinate the president's policy, constituted of today's equivalents of the Departments of Agriculture, Army, Commerce, Energy, Housing and Urban Development, Interior and Transportation, the Environmental Protection and Federal Emergency Management Agencies and the Tennessee Valley Authority. No major new water projects were authorized between 1970 and 1986.

Some observers have pointed out that while total damages have been increasing over time, damages relative to Gross National Product (GNP) have not. They argue that sufficient economic benefits have been achieved from flood control projects, in terms of increased land value and economic outputs, to more than justify the programs.[15] A 1992 assessment of floodplain management discusses this approach and concludes that "Overall relative [to GNP] flood damage appears to have remained, on the average, basically constant [from 1929 to 1983]."[16]

As the absolute flood damages increased, pressure had been put on the federal government, now officially in charge of preventing them, to make the victims whole. A new federal responsibility, potentially the costliest of all, was gaining ground: compensation for flood damages. Responsibility for the disaster relief acts enacted in the 1950s was distributed among a number of federal agencies but the Disaster Relief Act of 1974, enacted a century after the first federal disaster relief bill had been passed in 1874,[17] addressed the programs as a whole. In 1979 an entirely new agency, the Federal Emergency Management Agency (FEMA), was created by Executive Order 12127 to administer the disaster assistance, the NFIP, emergency management and other related programs. In 1988 this Act was amended, expanded and renamed the Stafford Act.

THE ENVIRONMENTAL DECADE

In the 1970s the federal government took on a major new responsibility, the protection of the environment. The National Environmental Policy Act of 1969 established the National Environmental Council and required all agencies to incorporate an assessment of environmental damages into the federal decision making process. This was followed, during the next ten years, by a series of tough new laws to protect the environment of which the most important, from the point of view of floodplain management, was the Clean Water Act, particularly Section 404 of the 1972 amendments to the Federal Pollution Control Act. The Clean Water Act established "water quality which provides for the protection and propagation of fish, shellfish and wildlife..." as a national goal, and Section 404 required the Corps to issue permits "for the discharge of dredged or fill material into the navigable waters at specified disposal sites" stating that disposal sites could be prohibited wherever it was determined that "the discharge of such materials into such area will have an unacceptable adverse effect on municipal water supplies, shellfish beds and fishery areas (including spawning and breeding areas), wildlife or recreational areas." The protection of aquatic ecosystems had become a national priority.

In 1973 the WRC adopted a set of *Principles and Standards for Planning of Water and Related Land Resources* (P&S) that required all federal actions to be evaluated in the context of four accounts: National Economic Development (NED), Environmental Quality, Regional Development and Social Well Being. It required that both monetary and nonmonetary effects be revealed in the accounts. The complexity and cost of such a process was enormous, and in response to a directive in the Water Resources Development Act of 1974 to study the P&S, an interagency study team produced a 22-volume report in 1975.[18] The 1974 Act also required consideration of nonstructural solutions to water resources projects and specifically authorized the acquisition of flood prone properties in three projects.

The illusion, created in the environmental decade of the 1970s, that there would be straight line progress away from economic protection and toward environmental management of the floodplain, did not materialize. After the WRC and the River Basin Commissions were abolished, the *Economic and Environmental Principles and Guidelines for Water and Related Land Resources for Implementation Studies* (P&G) were promulgated in 1983 to replace the old *Principles and Standards*. They not only allowed more flexibility but also relied entirely upon the NED at the expense of the Environmental Quality, Regional Development and Social Well Being accounts. Although the language called for plans to be "consistent with protection of the nation's environment," those plans ultimately had to be selected for their contribution to the NED as measured in dollars.

Responsibility for federal floodplain management activities had been extended beyond the U.S. Army Corps of Engineers. The *Unified National Program for Floodplain Management*, first issued in 1965 and amended in 1976, was again updated in 1979 and 1986 and the Water Resources Development Act passed in 1986 extended NFIP authority over all federally financed local protection projects and broadened local cost sharing for water resources projects. The Food Security Act of 1985 included "Swampbuster" provisions that provided farmers with incentives to avoid further drainage of wetlands.

By the beginning of the 1990s a number of new agencies and approaches were vying with the old dams and levees strategy of the Corps, but the conflicts and disproportionate responsibilities among the different levels of government, different agencies within the federal government and different outside groups having an interest in floodplain management were far from being resolved. In 1993 heavy flooding in the upper Mississippi River basin prompted several legislative actions: strengthening of the NFIP, tying Agricultural Disaster Relief to the Crop Insurance Program and creating an Emergency Wetland Reserve Program. The Floodplain Management Assessment published by FEMA in 1992, several post-flood documents generated by the federal Interagency Floodplain Management Task Force (FIFMTF), a Corps of Engineers assessment of floodplain management of the upper Mississippi and the National Research Council's analysis of flood control planning on the American River upstream from Sacramento, California all attempted to make sense of the potpourri of programs that today play a role in national floodplain management. This profusion of programs, implementing entities and policy documents demonstrates that flood control has moved well beyond the single purpose engineering thrust of its beginnings.

REFERENCES

1. Reuss M, Walker PK. Financing Water Resources Development: A Brief History. Fort Belvoir, Virginia: Office of History, US Army Corps of Engineers, 1983.
2. McCorvie MR, Lant CL. Drainage district formation and the loss of midwestern wetlands, 1850-1930 Agricultural History Vol 67, No. 4, 1993.
3. Arnold JL. The Evolution of the 1936 Flood Control Act. Fort Belvoir Virginia: USACOE Office of History, 1988:8.
4. *Ibid,* 22.
5. *Ibid,* 92.
6. Arnold JL. The flood control act of 1936: a study in politics, planning and ideology. In: Rosen H, Reuss M, eds. The Flood Control Challenge: Past, Present and Future. Proceedings of a National Symposium, New Orleans, Louisiana, September 26, 1986. Chicago: Public Works Historical Society, 1988:22.
7. Arnold JL. The Evolution of the 1936 Flood Control Act. Fort Belvoir Virginia: USACOE Office of History, 1988:94.
8. Steinberg B. Flood control in urban areas: past, present, and future. In: Rosen H, Reuss M, eds. The Flood Control Challenge: Past, Present and Future. Proceedings of a National Symposium, New Orleans, Louisiana, September 26, 1986. Chicago: Public Works Historical Society, 1988:90.
9. LR Johnston Associates. Floodplain Management in the United States: An Assessment Report, Volume 2. Federal Interagency Floodplain Management Task Force, 1992:2-28.
10. Reuss M. Comments. In: Rosen H, Reuss M, eds. *op cit,* 103.
11. Moore DP, Moore JW. The Army Corps of Engineers and the Evolution of Federal Floodplain Management Policy. Boulder: Institute of Behavioral Science, University of Colorado, 1989:35.
12. *Ibid,* 63.
13. Kates RW, Burton I, eds. Selected Writings of Gilbert White, Vol 1. Chicago: University of Chicago Press, 1986.
14. White GF. When may a post-audit teach lessons? In: Rosen and Reuss, eds. *op cit,* 56.

15. Shabman L. The benefits and costs of flood control: reflections on the flood control act of 1936. In: Rosen H, Reuss M, eds. *op cit,* 114.
16. Johnston, *op cit,* 3-36.
17. Schad TM. Evolution and future of Flood Control in the United States. In: Rosen H, Reuss M, eds. *op cit.*
18. Buie E. A History of Water Resource Activities of the US Department of Agriculture. Washington: USDA SCS, 1979, 24.

CHAPTER 5

THE GOVERNMENTS AND THEIR ROLES

States' rights prohibited a federal role in flood control until well into the 20th century but as the federal role has strengthened, the role of the states has diminished proportionately. Local governments have played a stronger role than states as partners in federal projects, but they have failed to play the role that other governments want them to: enforcer of floodplain regulations. Major watersheds are multistate. Rivers and floodplains cross and frequently define state lines. Watershed management usually requires a jurisdiction larger than a single state, but watersheds have rarely been planned or managed as complete entities. Flood control structures are relatively local in their impacts and specific in their benefits yet most of the costs of their construction are spread across the entire nation. Floodplain regulation is under the jurisdiction of the local governments, yet the burden of failure to implement is carried by the federal government, which hands out the disaster aid. The only truly national interest driving floodplain management programs is ecologic protection, whose benefits are sufficiently abstract and diffuse that they find a constituency only in the national population at large and rarely, if ever, in the interest groups that dominate local or state governments.

The level of government that should address a particular public problem, theoretically, depends upon the appropriateness of both its decision making capability and its tax base. Our floodplain management programs fall substantially short of meeting that criteria. The federal government's major role is the construction of projects that provide local benefits and distribution of disaster assistance payments to local beneficiaries. Federal flood control programs attempt to involve the local and environmental interests by consultation during the planning period, but they are basically planned and built by engineers. The National Flood Insurance Program is an attempt by the federal government to protect its own financial interests by coercing local governments to act against their own self interest. Federal investment in ecologic enhancement, for which it is particularly well suited, is minimal.

Federal floodplain management programs are designed to both prevent damages to the properties which abut our rivers and at the same time to protect and preserve the natural ecosystems which depend upon the floodplain for their existence. Dams and levees are built to prevent flooding from causing economic damage but their effects are to damage natural ecosystems. Programs which attempt to prevent new building on the floodplain and to flood proof existing buildings reduce economic damages, without adding to environmental ones. Flood victims who sustain damages are compensated by government aid programs. Floodplain acquisition programs both protect and restore the floodplain ecosystems and also insure an end to economic damages. The federal government plays the major role in implementing floodplain management programs but often depends upon state and local assistance in making them fully effective. The local authority to regulate floodplains continues to be the pivotal function upon which all else depends.

THE FEDERAL ROLE

Federal programs designed to manage activities on the floodplain fall into four categories:

- Structural programs that would reduce and avoid economic damages by preventing flood waters from reaching development on the floodplain. These are the structural projects, the levees and reservoirs built primarily by the U.S. Army Corps of Engineers (the Corps).
- Nonstructural programs that would reduce and avoid economic damages by preventing new damageable development from being built on the floodplain and by

reducing the potential for damage to the development that already exists. These are the regulatory and flood proofing requirements of the National Flood Insurance Program (NFIP), the floodplain buyout programs of the Federal Emergency Management Agency (FEMA) and the preparedness and warning programs of the National Weather Service (NWS) and the Corps.

- Aid programs that would compensate disaster victims for economic damages suffered from flooding. These are the disaster assistance and subsidized insurance programs administered by FEMA and the U.S. Department of Agriculture (USDA).
- Programs that would protect ecosystems by preserving and restoring natural floodplains. Acquisition programs of the Fish and Wildlife Service (FWS) and, to a certain extent, some programs implemented by the Corps and the USDA further this objective.

STRUCTURAL PROGRAMS

The Corps of Engineers is the lead agency in the construction of dams and levees. It was created in 1802 to design and construct fortifications. The engineering expertise of this branch of the Army was applied, during the 19th century, to the construction of navigation and flood works. Its congressional assignment, in 1927, to conduct the "308" watershed planning studies expanded and strengthened its hydrologic capability, establishing it as the appropriate federal agency to take the lead in the national flood control program initiated in 1936. In addition to implementation of major flood control works, the Corps has a continuing responsibility for repair and rehabilitation of flood control works, channel alternations, shoreline protection, flood fighting, emergency operations and floodplain information studies. An historian has characterized the engineering accomplishments of the Corps as "one of the largest single additions to the nation's physical plant—rivaled only by the highway system."[1]

The process of bringing a Corps structural project to completion is a long and careful one. It is usually initiated at the request of a local government and begins with first a reconnaissance study and then the execution of an elaborate feasibility study that evaluates the economics, prepares an environmental impact statement, elicits public opinion and secures a local sponsor. The project that is finally selected must receive congressional approval prior to actual construction.

The Natural Resources Conservation Service (NRCS), formerly the Soil Conservation Service of the USDA has constructed approximately 3,000 dams under its small watershed or "566" program and claims that by 1977 it had already been "involved in" more than two and a half million impoundments, including farm ponds.[2] Flood control structures have also been built by the Bureau of Reclamation, whose water-related operations are restricted to the 17 western states, and the Tennessee Valley Authority, which operates on the Tennessee and Mississippi Rivers.

NONSTRUCTURAL PROGRAMS

The NFIP was intended to be the kingpin in the federal initiative to prevent further encroachment onto the floodplain when the enabling legislation was passed in 1968. Today the NFIP is administered by FEMA and makes federally subsidized flood insurance available to property owners in local communities that themselves have joined the program. Although the NFIP has no authority over either local communities or their regulatory functions, it requires that a community perform certain damage avoidance activities before its residents can purchase flood insurance. Participating communities are required to enact floodplain regulations that will prevent new building in the floodplain and reduce damages to existing development. Building codes, permit procedures, subdivision and zoning regulations, flood proofing and elevation requirements are among the regulatory devices imposed by communities to prevent and reduce damages within the flood hazard areas identified by the NFIP. The NFIP has initiated a Community Rating System that rewards communities that adopt tougher floodplain management programs than those required by the program. The 100-year floodplain is the NFIP regulatory zone. Those areas have been defined in most of the participating communities by an extensive mapping program funded by the NFIP and carried out over the last 25 years. As of mid-1995, 18,203 communities were in the program and three billion policies were in effect.

The second thrust of the NFIP is to shift the costs of floodplain occupancy onto the property owner through the purchase of insurance, in order to eventually reduce the federal cost of disaster assistance to the amount of the subsidy invested in the program. Participation in the program has always been low, however. The NFIP estimates that only 25% of properties in flood hazard areas are covered by insurance policies (personal communication from the NFIP Administrator, 9/14/95). The only enforcing mechanism has been through federal lending institutions, which are expected to require flood insurance as a prerequisite for loans to property in flood hazard areas. Banks and savings and loans have often ignored this requirement and even when they complied initially, floodplain property owners were free to drop the insurance policies later without penalty. Reform legislation passed after the 1993 floods provides the NFIP with tools to strengthen lender compliance in Subtitle B; codifies the Community Rating System which rewards communities for good floodplain management in Subtitle C; mitigates flood risks in Subtitle D; and sets up Interagency Task Forces to deal with lender compliance and the Natural and Beneficial Functions of Floodplains in Subtitle E.[3]

Since 1988, FEMA has been able to allocate a portion of its disaster relief funds to the purchase and removal of damaged properties in the floodplain. Section 404 of FEMA's enabling legislation, the Stafford Act, authorizes the agency to contribute up to 50% of the cost of these hazard mitigation measures, using 10% of all FEMA grants for this purpose. Immediately after the 1993 floods Congress increased these percentages to 75 and 15, respectively, allowing federal agencies to eventually invest over $200 million to remove 8,251 damaged structures from the midwestern floodplains. FEMA contributed $105.6 billion; other federal agencies that provided funds were the Department of Housing and Urban Development through Community Development Block Grants ($67.1 million), the Economic Development Administration ($21.5), and the NFIP under its Section 1362 authority ($4.1 million).[4]

DISASTER AID

The Federal Emergency Management Agency (FEMA) coordinates most of the nonagricultural direct disaster assistance, in counties where a presidential disaster has been declared, through two programs. The Public Assistance program provides grants to local governments to compensate for damage to facilities and infrastructure; the Individual and Family Assistance Program provides direct aid to victims of the floods, including temporary housing, compensation for property loss and counseling. Over five billion dollars in disaster assistance was distributed by FEMA in 508 presidentially declared disasters between 1965 and 1989,[5] and $27.6 billion in 205 disasters between 1989 and 1993.[6]

The USDA provides direct compensation for damage to crops through two separate and sometimes overlapping programs funded by the Commodity Credit Corporation: the Crop Insurance Program and Agricultural Disaster Assistance. The Crop Insurance Program was created in 1938 and became a major assistance program in 1980. It first covered wheat, then cotton, and in 1948 it expanded to include sorghum, barley, corn and rice. By 1980 it covered 30 crops and by 1991, 51 crops. Participation in the program was 40% in 1990 and has declined since then. The program is heavily subsidized and it is estimated to have cost the government nearly $10 billion between 1985 and 1993.[7] During this same period another $9 billion was paid directly to farms in disaster assistance, though more for drought than for flooding. There were separate disaster assistance bills in 1987, 1988, 1989 and 1993 and disaster appropriations included in other bills in 1986, 1991 and 1992.[8]

Since passage of the 1994 Crop Insurance Reform legislation, a minimum level of crop insurance coverage has been required as a prerequisite for receiving disaster assistance. No premiums are charged for this level of coverage, although participating farmers must pay a handling fee of $50 per crop per county. The coverage guarantees the farmer a payment equivalent to 60% of the expected market price, for 50% of his average crop, according to the USDA Crop Insurance Program Fact Book. The reform legislation makes it more difficult for Congress to appropriate agricultural disaster funds, since crop insurance expenditures do not have the "emergency" status that exempted disaster assistance legislation from the "pay-as-you-go" restrictions that are used by Congress to rein in deficit spending.

ECOLOGIC PROTECTION

The most important federal programs for the protection of floodplain ecosystems from the damages incurred by structural flood control programs are the fish and wildlife habitat acquisition programs administered by the FWS. The "Pittman-Robertson" program restores wildlife habitat, the "Dingall-Johnson" program restores sport fish, and both programs provide 75% of the total project costs. "Partners for Wildlife" is a FWS program that provides grants to landowners to restore wetlands and riparian habitats. Several wetland protection programs administered by other agencies, such as the Corps' 404 permit program and the Swampbuster provisions of the Food Security Act of 1985, affect floodplain management peripherally, but their focus is not exclusively or even predominantly on floodplains. The USDA's Water Bank Program, the Department of Interior's Endangered Species Program, the Wild and Scenic Rivers Act and Coastal Zone Management are among other federal activities that potentially provide enhancements to floodplain ecosystems.

The USDA administers two buyout programs that both remove agricultural damages from the floodplain and add ecologic values to degraded former wetlands: the Wetlands Reserve Program (WRP) and the Emergency Wetland Reserve Program (EWRP). While the WRP may cover upland as well as floodplain wetlands, the EWRP applies exclusively to agricultural lands that have been so severely damaged by overbank flooding that it is not cost effective to rehabilitate them and the associated levees. Both programs purchase long-term easements on cropland that has been converted from former wetland, restoring it to its original state. Neither program makes much of an impact. The WRP was authorized in 1990 with a goal of restoring one million acres, but has restored only 39,000 acres, while the EWRP, approved in the fall of 1993, has approved applications covering 57,254 acres.[9]

THE PRICE OF FLOODPLAIN MANAGEMENT

Our major floodplain management expenses, to date, have been for the construction of flood control projects and for disaster assistance. Yet our figures for federal expenditures on flood control are no better than our data on any other aspect of floodplain management. The best overall number that FEMA was able to provide in its 1992 assessment of floodplain management was an old Water Resources Council estimate of more than $13 billion spent on flood control structures between 1936 and 1975. A more current number comes from Bory Steinberg, in 1986, who said that, so far, the Corps had invested $23 billion in flood control projects.[10] The FEMA assessment also estimated that the Corps flood control expenditures have been relatively stable at $1.1 billion per year since 1982.[11]

Public estimates of federal expenditures on disaster aid are no better. Neither the FEMA or Galloway Report numbers quoted above include agricultural aid, which is substantial. They do include disasters other than flooding, which is minimal. All we know is that we spent $5.5 billion per year on urban disaster assistance over the last five years and maybe an equal amount on agricultural disaster aid. Our disaster aid expenditures, therefore, are vastly exceeding the average annual $1.1 billion spent on construction (Fig. 5.1). The Stafford Act limits spending on the mitigation "buyout" projects to no more than 15% of all funds allocated to FEMA in special disaster legislation, and the mitigation funds available on a continuing basis are far less. Amounts spent removing properties from the floodplain, therefore, are best measured one disaster at a time, beginning with the $200 million spent in 1993.

Federal government assistance expenditures, in fact, may be exceeding the actual damages. In 1993, for example, when the Corps draft damage data suggested that total damages were not much more than $6 billion (see chapter 3), the major disaster relief legislation PL 103-75 allocated $5.7 billion for relief in August 1993 and an indeterminable amount thereafter. The SAST Report estimated that federal expenditures for agricultural damage alone, which represented about $2.5 billion of the PL 103-75 allocation, had reached $5.4 billion in June, 1994, and were "still increasing."[12]

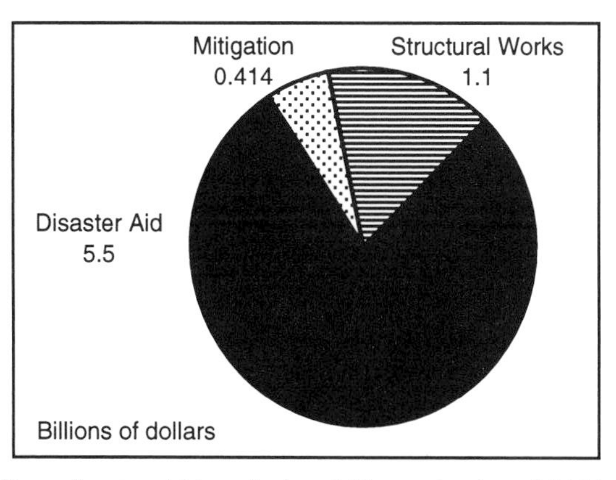

Fig. 5.1. Natural disaster costs, average annual expenditures. ***Disaster Aid:*** *The Galloway Report (Sharing the Challenge, 1994) estimates $27.6 billion was spent on disaster assistance programs from FY1989 through FY1993.* ***Structural Works:*** *The FEMA Assessment (Johnston, 1992) reports that annual outlays for Corps' flood control projects have stabilized at around $1.1 billion since 1982.* ***Mitigation:*** *The Environmental Working Group (Sowing Disaster, 1994) estimates that over half our disaster aid is agricultural. The projection of $414 million annually is 15% of the FEMA half of disaster aid.*

WHO PAYS?

The federal government has required financial assistance from local governments in the implementation of flood control projects for two reasons: first, to keep total costs down and second, to impose partial equity in a situation where most of the benefits were accruing to local property owners. Private levee districts were financed, of course, without federal assistance. In the 1917 Flood Control Act local interests paid one-third of the costs but in 1927 Congress elected to relieve the local government units on the lower Mississippi River of financial responsibility because of their inability to pay. One of the most contentious items in the 1936 Flood Control bill was the issue of cost sharing, eventually resolved in the so-called ABCs. Legislation in 1938 removed local cost sharing responsibilities from reservoir projects, according to Arnold,[13] so that the federal government could develop hydroelectric projects.

Cost sharing has gone through various iterations over time. Rubin reports that between 1917 and 1936 the average local share of costs for flood control works was 14%.[14] Rubin also points out that for this same period the federal government contributed, on the average, 93% to rural flood protection projects and 83% to rural ones. The federal contribution varied among the agencies, as well, with the Bureau of Reclamation paying 100% of the costs, the SCS paying 81% and the Corps, 95%. The 1986 Water Resources Development Act expanded the local financial responsibility to between 25–50%. The Galloway Report points out that

FEMA requirements for 25% cost sharing in mitigation and disaster assistance programs were reduced to 10% after the 1993 floods[15] and urges that the 25/75 local/federal ratio be maintained in all FEMA programs. The fact is, in the aftermath of the 1993 floods, other federal agencies such as the Department of Housing and Urban Development sometimes paid the local government share. Whatever cost sharing arrangements are agreed upon, local governments are always short of funds and frequently successful in avoiding them.

THE ROLE OF THE STATES

States frequently play a necessary but rarely a sufficient role in floodplain management. They supplement federal programs and facilitate local ones with education, information, technical assistance, regional and watershed planning mechanisms and sometimes grants. The degree to which they are effective varies according to the commitment, the resources and the flooding characteristics of the individual states.

The best source of data we have on state floodplain management programs comes from the triennial surveys of state and local programs conducted by the Association of State Floodplain Managers.[16] According to the 1992 survey, 32 states construct or participate in the construction of structural projects and 28 states provide full or partial funding for structural projects. Twenty-three states regulate new dam construction and only seven states and territories do not have dam inspection programs. Nineteen states regulate levee construction.

Survey data shows that all states regulate their own development activities on the floodplain, yet many of them exempt economically important activities from those regulations. States participate in a variety of activities in support of local governments in their implementation of NFIP regulatory requirements and many states even impose floodplain regulation or building standards that are stricter than the federal ones. Most states provide information about flood proofing and emergency preparedness and some states operate flood warning systems. Some provide tax incentives to flood prone property owners to keep their land undeveloped.

Although most disaster aid is provided by the federal government, the survey indicates that most states encourage property owners to buy flood insurance, and almost all states provide flood emergency assistance, contribute to the nonfederal share of federal disaster assistance and assist local communities with post-disaster recovery. A total of 22 states had been recently involved in multi-objective management activities, and 43 states regulate wetlands in some fashion, according to the survey. Many states participate in federal programs such as Coastal Zone Management, Wild and Scenic Rivers and Threatened and Endangered Species.

The state role in floodplain management is diffuse, yet even the Galloway Report, which stresses the need to resolve problems of intergovernmental role confusion, seems to rely upon state membership on basin commissions to make the difference. The specific Galloway proposals for new state initiatives are not very substantial: that states regulate levees that don't fall under the Corps program; that states should actively encourage flood insurance purchase; that states should shoulder certain local cost share; that state floodplain management officials should encourage school districts to include natural hazard in curricula; and that the states should assume an increased role in all floodplain management activities.[17] If there is an independent and effective role for states in floodplain management, it has not yet been defined.

State interest in floodplain management, as might be expected, is strongest within the staffs of agencies assigned to implement the floodplain management tasks. In a 1983 survey of all 50 state governments conducted by Burby and Kaiser, while 96% of agency personnel say flooding is a serious problem, only 40% of the state legislators saw it in the same light. A related survey demonstrated, however, that states rank flooding as a higher priority than do local government governments. Out of a sample of 956, only 17% of local government officials rated flooding as a serious problem.[18]

THE LOCAL ROLE

While local communities sometimes build small flood control and storm water management structures and usually provide post-flood assistance to the extent that they are able, their primary role in the management of floodplains is to regulate their use. The local community, in fact, is the unit of government that has the main authority to regulate development on the floodplain. The federal government provides incentives for regulation and state governments provide encouragement, technical assistance and training to local floodplain managers, but the regulation itself depends upon the intent and the ability of the local officials to accomplish it. The tools they have available to them to restrict new development on floodplains are zoning and subdivision regulations, building and housing codes, flood proofing and elevation requirements. County governments often provide property tax relief, as either compensation for flood damages or incentives for nondevelopment. Local interests are required to share the costs of many federal programs. It is not uncommon, however, for state or even other federal agencies to provide the funds for local cost sharing.

Drainage and levee districts, strictly classified as local units of government but in fact organizations of private property owners within the geographical area protected by a given levee, usually provide local sponsorship for Corps structural projects. In this capacity they provide a share of the costs (normally 25%), are responsible for operations and maintenances, and provide the rights of way on which the project is built. Financing for these activities is acquired by special drainage taxes imposed upon the district members, according to the degree of flood protection they receive.

The Galloway Report assumes that the NFIP is the proper channel for federal expectations of local government performance to be fulfilled. The report proposes one action which illustrates the ambivalence inherent in federal recommendations for local action: "encourage communities to obtain private affordable insurance as a prerequisite to receiving FEMA public assistance money." It is not clear whether *encourage* or *prerequisite* is the operative word.

It began as the business of private property owners and eventually grew to be a major and almost the exclusive responsibility of the federal government. Floodplain management is talked about today as an intergovernmental responsibility, but there is no good match between the funders and the beneficiaries; between the planners and the invested interest groups. Federal initiatives require intense interagency cooperation and federal program goals cannot be met without major state and local efforts, yet we are looking for better ways to accomplish both. Today we have a wide variety of floodplain management programs, yet for fifty years we poured money and effort into holding back the floodwater with structural constraints and it could be argued that the system is still driven, in a negative sense, by the levee and reservoir programs of the Corps of Engineers. That structural program, for all the problems it has solved, has caused some new ones of its own that need to be addressed.

REFERENCES

1. Arnold, JL. The Evolution of the 1936 Flood Control Act. Fort Belvoir Virginia: USACOE Office of History, 1988:91.
2. LR Johnston Associates. Floodplain Management in the United States: An Assessment Report, Volume 2. Federal Interagency Floodplain Management Task Force, 1992:12-9.
3. National Flood Insurance Program. Watermark, Spring and Summer 1995. Washington DC: Federal Emergency Management Agency, 1.
4. United States Army Corps of Engineers. Floodplain Management Assessment of the Upper Mississippi River and Lower Missouri Rivers and Tributaries. 1995:7-9.
5. Johnston, *op cit,* 3-15.
6. Interagency Floodplain Management Review Committee. Sharing the Challenge: Floodplain Management into the 21st Century. Report to the Administration Floodplain Management Task Force, 1994:181.
7. Hoffman WL, Campbell C, Cook KA. Sowing Disaster. The Implications of Farm Disaster Programs for Taxpayers and the Environment. Washington DC: Environmental Working Group, 1994:10-11.
8. *Ibid,* 17-18.
9. United States Army Corps of Engineers. Floodplain Management Assessment of the Upper Mississippi and Lower Missouri Rivers and Tributaries.1995:7-19.
10. Steinberg B. Flood control in urban areas: past, present, and future. In: Rosen H, Reuss M, eds. The Flood Control Challenge: Past, Present and Future. Proceedings of a National Symposium, New Orleans, Louisiana, September 26, 1986. Chicago: Public Works Historical Society, 1988:90.
11. Johnston, *op cit,* 12-2,3.

12. United States Army Corps of Engineers. Floodplain Management Assessment of the Upper Mississippi River and Lower Missouri Rivers and Tributaries. 1995:191.
13. Arnold, *op cit,* 74.
14. Rubin K. The new federalism and national flood contol programs. In: Rosen H, Reuss M, eds. *op cit,* 126-127.
15. Interagency Floodplain Management Review Committee. *op cit,* 82.
16. Association of State Floodplain Managers. Floodplain Management 1992: State and Local Programs. Madison, Wisconsin: Association of State Floodplain Managers, 1993.
17. Interagency Floodplain Management Review Committee. *op cit,* 81,102,108,111,133.
18. Burby RJ, Kaiser EJ. An Assessment of Urban Floodplain Management in the United States: The Case for Land Acquisition in Comprehensive Floodplain Management. Madison, Wisconsin: Association of State Flood Plain Managers Technical Report #1, 1987:7.

CHAPTER 6

HARMFUL EFFECTS OF FLOODPLAIN MANAGEMENT

THE PROBLEM

Throughout the 19th century and well into the 20th, if floodplains were managed at all it was by building structures to control the floods. Floodplain management began by putting development on the floodplain and ended by raising a levee to protect it from the annual floods. Later, reservoirs were constructed, upstream, to hold back the waters that otherwise would cover downstream floodplains. Often the channels were cleared, widened and deepened or new channels were dredged to hold more water when it flooded.

These flood control measures, themselves, created other problems. First, the construction of barriers to prevent the water from reaching the floodplain endangered and destroyed the fragile ecosystems which depend upon this critical link between the upper watershed and river channel. Next, the barriers attracted new developments that, in times of extreme flooding, resulted in even greater damages. Today floodplains must be managed to correct or compensate for these new damages caused by the flood control works.

ECONOMIC IMPACTS OF FLOOD CONTROL

The positive impacts of flood control projects, their benefits, are the damages that they prevent. To determine the benefits of any project, the potential damages that would occur without the project are counted. Federal laws, as we have seen in chapter 4, require flood control measures to demonstrate positive benefit/cost (B/C) ratios, namely that the predicted annual benefits exceed the construction costs of the project. The costs are largely borne by the federal government. The benefits are, most often, reaped by individual land owners.

Benefits are typically calculated, in designing flood protection, by first identifying at least two kinds of damages that will be prevented by a particular level of protection: property damages and loss of agricultural income. The reach of river subject to the flooding is described hydrologically by a discharge-frequency curve that relates peak flows (discharges) to different probabilities of occurrence. This relationship ideally is developed from a long sequence of historical records, but sometimes it is derived from comparison with other watersheds, or even from estimated runoff from precipitation.[1] The size of the discharge is then used as the basis for calculating the associated stage of the flow, by taking into consideration such things as channel capacity and roughness. The calculated stage, in turn, is the basis for determining the amount of inundation and therefore damage that will occur. The frequency of occurrence of that stage indicates how often floods of that magnitude will occur. On the discharge-frequency curve, for example, the 100-year flood represents a peak discharge and a volume of water that will overflow the river banks and inundate the floodplain along that reach up to a certain elevation. The benefits that are achieved by building a levee to contain that 100-year flood are said to be equal to the damages that such a flood and other smaller floods would cause to homes and buildings below that elevation when the levee is not present.

If, instead of being covered by residential development, the entire floodplain was normally planted with corn, the benefits of the project would be equal to the loss of net profits the farmers would suffer when floods of different frequencies and peak flows (defined, again, on the discharge-frequency curve) wiped out their crops. The actual agricultural benefits are calculated by a formula that determines the production value of acreage behind the levee:

a standardized dollar per acre value is given to whatever crops are typically grown in a particular region. The damages are then annualized over the projected life of the levee and averaged out so they can be compared with costs. The costs are simply construction estimates. Their annualization, however, requires the selection of a discount rate in the calculation of future value, which affects the magnitude of the final number.

The calculation of a B/C ratio for a protection project, therefore, roughly involves the following steps:

- Collection of historical data on the frequency of flows (discharges) of different sizes;
- Generation of a discharge-frequency curve for the reach of stream;
- Determination of the stage-discharge relationship;
- Mapping of the floodplain on that particular reach to determine what area would be flooded by the stages associated with floods of the design frequency or smaller;
- Calculation of damages produced by inundation to that elevation;
- Annualization of those damages over 30 years. These are the benefits;
- Estimation of the construction costs;
- Selection of a discount rate and annualization of costs over 30 years. These are the costs; and
- Division of the annualized benefits by the annualized costs. This is the B/C ratio.

If the B/C ratio is greater than 1.0, it is concluded that the benefits exceed the costs and the project is eligible for public funding.

Analytic models are used to understand and predict the hydrologic and hydraulic characteristics of river basins. Different agencies use different models and there is not always total consistency, even among individual U.S. Army Corps of Engineers (Corps) districts. Although a model called HEC-2 has been most commonly used in the past, it is characterized as a "steady-state, one-dimensional, rigid-boundary model," it has been criticized for being unable to account for levee breaches and storage effects,[2] and it has been replaced in recent studies by an "unsteady flow" model, the UNET.

Damage estimates, particularly of property, quickly become obsolete. The construction of a flood protection structure usually attracts new and more valuable development in the protected area. Homes and businesses move in behind a levee increasing, frequently dramatically, the damages being prevented by the structure. An issue surrounding B/C ratio calculation, therefore, is whether future benefits should be calculated and included. The Soil Conservation Service has been criticized for anticipating the increases in land values which would result from the construction of their small watershed projects (see chapter 4) and including those potential benefits in the calculation of the B/C ratio for a proposed project. If these projects guaranteed protection from all future flooding and if it were possible to accurately predict the future benefits, perhaps an argument could be made for the practice of counting them in project benefits. The opposite, however, is the case.

The expansion of development on floodplains behind flood protection structures may increase the benefits of those structures when a flood equivalent to or smaller than the design flood occurs, but in larger floods when levees are overtopped, it results in damages far greater than would have occurred before the protection was in place. These *residual damages* occur even when a protection structure is performing properly. The B/C ratio requirement assures that each flood control structure is designed to provide a level of protection that is justified by the damages it prevents. If 50-year is determined to be the most cost effective level of protection for an agricultural floodplain or a housing development, it stands to reason that the higher construction costs of 100-year or even 500-year protection could not be justified by existing damages. Those properties are expected to flood, therefore, in larger floods, regardless of how much new development has been constructed. Because a levee is overtopped, it is not an indication of a failure; it is more likely to be performing exactly as it was designed to do. Sandbagging and other flood fighting procedures are not anticipated in designs, distort the proper functioning of the levee and frequently do more harm than good from an engineering point of view.

Flood control projects reduce but never eliminate all risk from flooding. Once a project is in place, however, the public perceives (at least it acts as if it does) that all risk has been removed, often with good reason. In Johnstown, Pennsylvania the Corps District engineer announced when the flood control project was completed in 1943 that "the flood troubles of the city of Johnstown are at an end"[3] but they were not, as the devastating 1977 flash flood demonstrated. Likewise the "Guttenberg Report," produced by the Corps to support the upper Mississippi River valley projects that were authorized in the Flood Control Act of 1954, promised that "accomplishment of [the authorized projects]... would eliminate in large measure the damages in that part of the Mississippi River valley located within the Rock Island Engineer District,"[4] yet the Rock Island district report on damages in the 1993 floods makes note of the floods of 1965, 1969 and 1972 and says that "the flooding of the Mississippi River was the most devastating in terms of property loss, disrupted business and personal trauma of any flood in the history of the United States."[5] The engineers that build the projects and the politicians that fund them never talk about their limitations, so it is hardly surprising that the public believes all risk has been removed. New development occurs, protected values increase and with it, by definition, the damages when it floods. The true economic impacts of flood control projects, therefore, are much greater than the actual construction costs. They include the damages later incurred from new development that is stimulated by those projects.

There is another economic cost associated with flood control projects. Since rivers are part of continuous systems, the manipulation of the flows within them will produce effects elsewhere. Structural flood control projects are designed to manipulate the flows. The hydraulic effect of levees is to raise flood crests elsewhere. Construction of reservoirs to prevent inundation downstream increases inundation upstream. In their 1995 floodplain management assessment the Corps of Engineers calculated that the effect of raising the agricultural levees on the upper and middle Mississippi River to a level that would have prevented overtopping in 1993 would be to increase flood stages on the middle Mississippi by approximately six feet.[6]

ECOLOGIC IMPACTS OF FLOOD CONTROL

Construction projects that alter our river systems damage and destroy the natural ecosystems contained within them. Such projects may be designed for flood control, water supply, navigation, hydroelectric power, recreation or combinations of them all.

The National Research Council (NRC) Report on Aquatic Ecosystems provides a table from Karr et al, 1986 identifying five classes of environmental factors affecting aquatic biota that are disrupted by manmade alterations:[7]

- Energy sources: particulate matter entering the stream is affected in size, type and amount and the seasonal pattern is disrupted.
- Water quality: changes are made in temperature, turbidity, dissolved oxygen, nutrients, organic and inorganic chemicals, heavy metals and toxic substances and pH.
- Habitat quality: substrate type, water depth and current velocity, spawning, nursery and hiding places and diversity of habitat are affected.
- Flow regime: volume of water and distribution of high and low flows is disrupted.
- Biotic interactions: equilibrium in competition, predation, disease and parasitism is disturbed.

Flood control projects drastically alter the channel conditions. Free flowing waters are turned into a series of navigation pools or they are speeded up by straightening, channelization, removal of snags or diversion. Geomorphological changes occur as a result of bank erosion, bed scour and the destruction of stability in riparian soils. The canyons of the upper American River described in chapter 9 provide a good example.

The reservoirs that have permanently inundated millions of upstream acres have destroyed entire ecosystems and have produced downstream impacts that include blocking the migrations of aquatic organisms and disruption of the natural sedimentation rates. As deltas and sand bars subside, fisheries are destroyed. Dams on two rivers as different as the Mississippi and the Nile have destroyed important fisheries in their receiving seas, the Gulf of Mexico and the Mediterranean.[8]

Structures that deprive floodplains of the annual flood pulse and place barriers between the river channel and the upper watershed, however, generate the most damage to the overall health of riparian ecosystems, as described in chapter 3, and that is precisely what flood control projects are designed to do. Flood control projects remove water from the floodplain; ecosystems require its presence there.

It's difficult to know how much riparian habitat we have destroyed. The NRC report quotes several estimates of altered or lost natural stream bank habitat as a result of channelization activities, which range from 6% to 70%.[9] While we know that flood control projects change the way we use our floodplains we do not know how much we've lost and we're not sure how much we have to lose. It is one of many of the ecologic parameters of floodplain management that we are unable to quantify. The survey done by the Association of State Floodplain Managers asks respondents to assess the status of their states' floodplain resources. In the 1995 survey the average respondent indicated that his state was "holding steady" in aquatic habitat, riverine access and recreational opportunities, but that resources were (ever so slightly) "being lost" in riparian habitat, open space and inland wetlands.[10]

Because there is no way to reliably predict, much less measure, specific future ecologic damages from a flood control project, they have no impact on the calculation of a B/C ratio. For reasons discussed above in chapter 3, measurement is difficult and the monetization of actual ecologic damages is almost impossible. Predicting the long-term effects of disruption on fragile ecosystems presents even more problems. In the American River valley, where there is no documented record of the effects of inundation on nonriparian chaparral, digger pine, and oak woodland communities, for example,[11] there is no scientific basis for projecting the impact of the proposed project and attempting to assign a dollar value to it makes no sense.

The effects of a flood control project extend far beyond its immediate impact on the surrounding environment. We need to know what the long-range impacts will be and to decide if a specific project can be held accountable for future ecologic damages incurred when the inevitable new development is attracted to riparian locations, new infrastructure is installed and a whole sequence of changes occurs. We need to learn much more about the impacts of flood control projects on the riparian or floodplain environment.

Understanding the impacts makes sense, but monetizing them does not. Yet as long as we depend upon the B/C ratio to justify our projects, attempts to monetize will continue to be made. The system we have designed to justify activities on our floodplains and in our river systems is able to recognize $100 worth of property damage or a $1000 crop loss in a B/C ratio calculation, but cannot account for the destruction of an irreplaceable ecosystem. The best that we can do is to calculate the economic B/C ratio, describe the ecologic impacts and make a judgment call. Yet there is no place in the planning process for such a call to be made, nor authority given to the people who could make it best.

The B/C ratio that drives our structural flood control program is severely flawed in two important respects. It does not count among a project's costs either the increase in residual economic damages that are a direct result of the construction or the very substantial ecologic damages that result. Even the calculation of present benefits, as we will see in chapter 8, is subject to many uncertainties. These damages occur, however, over and over again. The residual effects of flood control projects certainly contribute to and may even be responsible for the continuing escalation of damages and costs which haunt our floodplain management programs. The ecologic damages have been severe, yet until the 1970s our policies barely took them into account. In the last 30 years, however, changes in national flood control policy have attempted to correct these two problems by a greater emphasis on nonstructural solutions.

References

1. LR Johnston Associates. Floodplain Management in the United States: An Assessment Report, Volume 2. Federal Interagency Floodplain Management Task Force, 1992:6-10.
2. Interagency Floodplain Management Review Committee. Sharing the Challenge: Floodplain Management into the 21st Century. Report to the Administration Floodplain Management Task Force, 1994:157.

3. Reuss M. Introduction. In: Rosen H, Reuss M, eds. The Flood Control Challenge: Past, Present and Future. Proceedings of a National Symposium, New Orleans, Louisiana, September 26, 1986. Chicago: Public Works Historical Society, 1988: xiii.
4. US House of Representatives Document #281.
5. United States Army Corps of Engineers. The Great Flood of 1993 Post-Flood Report: Upper Mississippi River and Lower Missouri River Basins. Appendix B. Chicago: North Central Division, Corps of Engineers, 1994:15.
6. United States Army Corps of Engineers. Floodplain Management Assessment of the Upper Mississippi River and Lower Missouri Rivers and Tributaries, 1995:4.
7. National Research Council. Restoration of Aquatic Ecosystems. Washington, DC: National Academy Press, 1992:189.

7a. The information is attributed to Karr JR et al. Assessing biological integrity in running waters: A method and its rationale. Special Publication 5. Illinois Natural History Survey, Champaign, Ill.

8. National Research Council, *op cit,* 177, 201.
9. *Ibid*, 194.
10. Association of State Floodplain Managers. Floodplain Management 1995: State and Local Programs. Madison, Wisconsin: Association of State Floodplain Managers, 1996. (In press)
11. National Research Council. Flood Risk Management and the American River Basin; An Evaluation, Washington DC: National Academy Press, 1995:92.

CHAPTER 7

THE THEORY OF FLOODPLAIN MANAGEMENT

CHANGING CONCEPTS

People were thinking and talking about comprehensive floodplain management long before they were doing it. For a century and a half they concentrated, in this country, on just keeping water off the floodplain by building levees. But even in the 19th century men with a broader vision were speaking out. Charles S. Ellet, Jr. wrote in a report to Congress in 1852 that "the greater frequency and more alarming character of the floods [on the Mississippi] are attributed *primarily*, to the extension of cultivation... *secondly,* to the extension of the levees..." These ideas of Ellet's, and others, were pushed aside in the 1860s after the publication of the more popular Humphreys and Abbot proposals for levee construction alone. A century later, however, after many more voices had spoken out, official policy began to consider a set of management tools that included not only structural flood prevention projects but also "nonstructural" damage prevention techniques such as flood proofing and restrictions on the uses of the floodplain. The goals of floodplain management policy eventually expanded to include prevention of ecologic as well as economic damages.

During the transformation of flood control into floodplain management the nation had shifted in the way it answered three questions:

- Why should floodplains be managed?

- Who should manage floodplains?
- How should floodplains be managed?

Floodplain management policy has been driven forward, in fits and starts, by major flood events (see Table 7.1 and Fig. 7.1). Appropriately, the 1993 major flooding in the upper Mississippi River basin produced an important review of national floodplain management policy, the 1994 *Sharing the Challenge: Floodplain Management into the 21st Century. Report to the Administration Floodplain Management Task Force* (The Galloway Report) and the accompanying *Science for Floodplain Management into the 21st Century*, followed by the transmittal of an update of the *Unified National Program for Floodplain Management to Congress* in 1995. Today's floodplain management theories are well represented and described in these documents.

SHARING THE CHALLENGE

The Galloway Report contains eleven conclusions related to the 1993 flood and 35 recommendations and 60 proposed actions directed toward overall improvements in floodplain management. The difference between recommendations and actions is that recommendations can be implemented without new resources while actions cannot. Six chapters and more than a third of the report is devoted to what is titled "A Blueprint for the Future."

The Galloway Report opens with an identification of the three problems faced by floodplain management today:[1]

- People and property on the floodplain still remain at risk;
- The "severe ecological consequences" of loss of habitat; and
- An unclear division of responsibility among federal, state, tribal and local governments.

The Report endorses the dual goals of damage reduction and ecologic protection. It says that we must "reduce the vulnerability... to the dangers and damages that result from floods" and "preserve and enhance the natural resources and functions of the floodplains,"[2] and it claims this can be done: "Floodplain resources can be shared by human occupants and natural systems."[3]

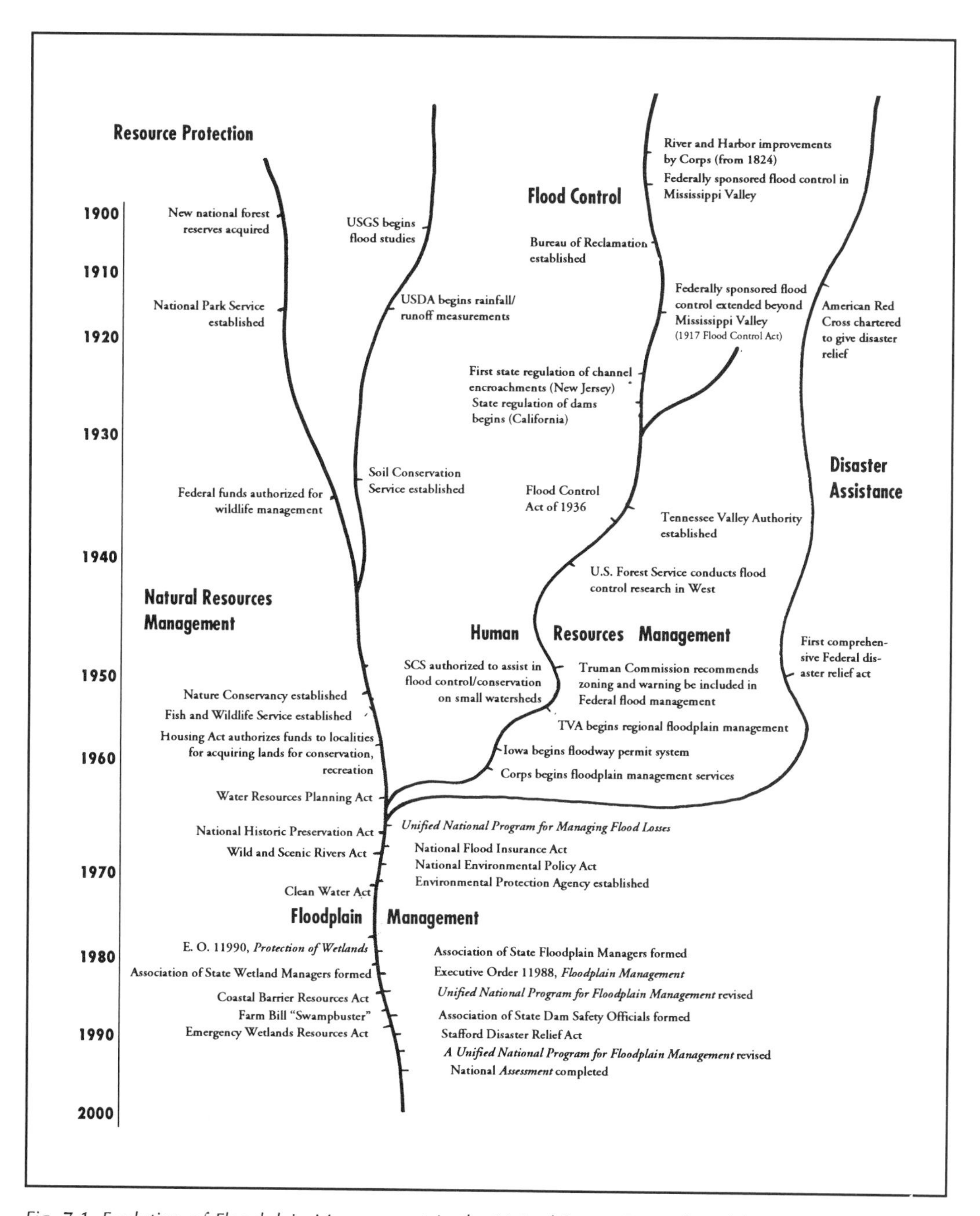

Fig. 7.1. Evolution of Floodplain Management in the United States. Reproduced from a Unified National Program for Floodplain Management 1994. Federal Interagency Floodplain Management Task Force.

Table 7.1. Some steps in the development of federal floodplain management policy

1849-1850	The Swamp Land Acts granted floodplain lands to the states to be drained and reclaimed for cropland.
1852	A report by Charles Ellet Jr. blamed flood damages on floodplain development, recommending reservoirs.
1861	*A Report on the Physics and Hydraulics of the Mississippi River,* by Captain Andrew A. Humphreys and Henry L. Abbot concluded that levees were the only effective flood control measure.
1874	The first congressional approval of "relief" for flooding.
1879	Creation of the Mississippi River Commission, which embarked upon a levees-only program.
1917	The Federal Flood Control Act of 1917 funded projects on the Mississippi and Sacramento Rivers, requiring local interests to pay one third of the costs.
1927	In the Rivers and Harbors Act the Corps was authorized to do comprehensive "308" basin surveys.
1928	The flood control work on the Mississippi was extended beyond levees (for the first time) to include floodways, spillways and channel improvements, with no local cost sharing.
1936	The Flood Control Act of 1936 established a national responsibility for flood control, spelled out the local ABC obligations; required a positive benefit/cost ratio, authorized over 200 projects, assigned responsibility for flood control to the Corps of Engineers and was amended to include a role for the US Department of Agriculture.
1938	In the Flood Control Act of 1938 local cost sharing requirements were eliminated for reservoirs to allow the federal government to retain title and control hydroelectric projects; the Corps was given authority, for the first time, to evacuate and relocate people from floodplains.
1944	The Flood Control Act of 1944 authorized the USDA to do flood protection in eleven watersheds.
1954	The Watershed Protection and Flood Prevention Act of 1954 authorized the USDA to implement a "Small Watershed Program."
1965	The Water Resources Planning Act of 1965 (PL 89-80) created the Water Resources Council (WRC).
1966	Bureau of the Budget Task Force on Federal Flood Control Policy recommended *A Unified National Program for Managing Flood Losses* (House Document 465), including flood insurance, to reduce losses.
1966	Executive Order 11296 directed federal agencies to evaluate flood hazard potential before locating new buildings in the floodplain.

Table 7.1. (continued)

1968	Passage of the National Flood Insurance Act.
1973	The Water Resources Council adopts *Principles and Standards for Planning of Water and Related Land Resources* which required considerations of regional, environmental and social well being impacts.
1975	Establishment of the Federal Interagency Floodplain Management Task Force (FIFMTF) within the WRC. Membership: Departments of Agriculture, Army, Commerce, Energy, Housing & Urban Development, Interior and Transportation; the USEPA and the Tennessee Valley Authority.
1976	Preparation of *A Unified Program for Floodplain Management.*
1979	Preparation and submittal to Congress of *A Unified Program for Floodplain Management.*
1979	The Federal Emergency Management Agency (FEMA) is established by Executive Order 12127.
1982	The WRC is disbanded and responsibility for the Unified Program is reassigned to FEMA, whose director assumes FIFMTF Chairmanship.
1986	Preparation and submittal to Congress of an updated *Unified Program for Floodplain Management.*
1986	The Water Resources Development Act of 1986 in which non-structural solutions are given more importance and local cost sharing is increased.
1988	The Disaster Relief Act of 1974 was retitled the Stafford Act and amended to guide flood recovery and mitigation practices.
1989	*A Status Report on the Nation's Floodplain Management Activity: An Interim Report,* was prepared by the FIFMTF.
1989	*Action Agenda for Managing the Nation's Floodplains,* a response to the *Status Report,* was issued by the National Review Committee.
1992	*Floodplain Management in the United States: An Assessment Report* was published by the FIFMTF.
1994	*Sharing the Challenge:Floodplain Management into the 21st Century* was prepared by a Floodplain Management Review Committee created by the FIFMTF to respond to the 1993 floods in the Upper Mississippi Basin and chaired by General Gerald E. Galloway.
1995	*Floodplain Management Assessment of the Upper Mississippi River and Lower Missouri Rivers and Tributaries* was prepared by the Corps in response to a congressional directive.
1995	*A Unified National Program for Floodplain Management, 1994* was prepared and submitted to Congress by President Clinton in March.

To accomplish these goals, according to the Galloway Report, we must:

- Avoid the risks of the floodplain;
- Minimize the impacts of those risks;
- Mitigate the impacts of damages when they occur; and
- Accomplish the above in a manner that concurrently protects and enhances the natural environment.

The Galloway Report proposes to clarify the roles of various levels of government by passing a Floodplain Management Act, activating the old Water Resources Council and establishing Basin Commissions to coordinate the newly defined roles. The Council would also coordinate the activities of federal agencies and insure that they set an example for good floodplain management. It recommends that the states take a leadership role in all floodplain management activities. It states outright that "Ultimate responsibility for floodplain management rests with individuals and local government..."[4] and urges that cost sharing formulas of 75% federal and 25% local be applied to all disaster programs. It suggests, however, that local governments be helped with their funding share of the environmental restoration and enhancement programs in Sections 906 and 1135 of the 1986 Water Resources Development Act.

Proposals in the Galloway Report for strategies to accomplish the goals are directed more toward improving present programs than creating new ones. The "(t)wo fundamental strategies—protection or removal"[5] are identified as the ways to minimize the impacts of flooding. The report proposes that the U.S. Army Corps of Engineers (Corps) be designated as the "principal federal levee construction agency"[6] and urges consistency in the cost sharing formulas of all federal agencies. It calls for administration support for Corps criteria for levee repairs, suggests that they be consistent with the levee repair criteria of the Natural Resources Conservation Service of the Department of Agriculture, and proposes that limits be put on flood fighting activities. The report also proposes that all population centers and critical infrastructure be protected from the standard project flood discharge (normally the 500-year flood). The strategy of "removal" is basically the federal buyout of damaged floodplain properties. The report proposes interagency cooperation and increased flexibility and funding in the existing hazard mitigation programs and specifically directs that a program be developed to reduce losses to repetitively damaged insured properties.

The report recommends that disaster recovery activities be strengthened and better organized. It argues that the purchase of flood insurance policies could be increased by more active encouragement of the states, improved lender compliance, escrow of premiums by lenders, and improved marketing techniques. It recommends reducing post-disaster support to those eligible for flood insurance and at the same time says that a safety net should be provided for low-income flood victims. It calls for extending the National Flood Insurance Program (NFIP) to all areas behind levees that are less than the standard project flood design and for extension of the 5 day waiting period for NFIP policy coverage to at least 15 days. It recommends limiting assistance to communities that are not in the NFIP and encouraging communities to get private insurance in order to be eligible for federal disaster assistance. It urges the NFIP to promote the Community Rating System, to complete its mapping of flood prone communities and to improve the accuracy and timeliness of those maps. Finally, the report says that the federal crop insurance program should be reformed by limiting the assistance it provides, increasing participating and making it more actuarially sound.

Accomplishing these things in ways that protect and enhance the environment, the Galloway Report suggests, requires multiobjective comprehensive watershed management, better organization and more funding for wildlife land acquisition programs. It recommends that both an ecological needs investigation and an ecosystem management demonstration project be initiated in the basin.

Most of the actions and recommendations in the Galloway Report correct inefficiencies in existing programs. The 31 members of the Review Committee that prepared the report were drawn from the Departments of Agriculture, Army and Interior, the Environmental Protection and Federal Emergency Management Agencies and the Office of Management and Budget. These professionals were able to successfully identify specific and often important organizational and operational deficiencies. When the report tackled the problem of environmental enhancement, for example, the one recommendation is to increase funding for the Refuge Revenue Sharing Act and the eight proposed actions are to establish a lead agency for federal land acquisition for environmental purposes; set up emergency implementation procedures for environmental land acquisitions; coordinate the land restoration activities of federal agencies; make the use of programmed funds for emergencies more flexible; focus acquisition efforts on sites with high natural values; include ecosystem management funds in operations and management budgets; help locals with cost sharing for environmentally positive alternatives to PL 84-99 (the Corps levee repair program) and fund mitigation at the same rate as construction. These may be necessary and valuable steps to take in the interest of greater program efficacy but whether or not these programs would then move us significantly faster toward ecosystem protection is quite another matter.

The Galloway Report recognizes the bias of the 1983 *Principles and Guidelines* against nonstructural measures and points out that the emphasis on monetary accounts precludes their effectiveness in addressing environmental issues. The report proposes actions that would reestablish the importance of environmental factors in planning, closer to the model that had been proposed by the 1973 *Principles and Standards.* These documents were discussed in chapter 4.

The Galloway Report calls for better data collection and more research. It recognizes the deficiencies in damage data collection and proposes a national inventory of flood prone structures. It recommends research to address the questions of why the NFIP has poor market penetration, what the impact of federal farm programs is on floodplain land use and the current system of funding disaster relief. It calls for the development of techniques to assess and monetize environmental and social benefits. It asks for reviews of the stream gauging network and flood forecasting and of discharge-frequency relationships. The Galloway Report recommendations that affect the states and local governments were discussed in chapter 5.

SCIENCE FOR FLOODPLAIN MANAGEMENT

A Scientific Assessment and Strategy Team (SAST) was convened under the chairmanship of John Kemelis of the United States Geological Survey in November 1993, again in response to the Mississippi floods. It was subsequently attached to the Galloway Committee, when the latter was formed in January 1994. The SAST defined its own scope of work to include a comprehensive data collection and analysis effort. The report it produced, *Science for Floodplain Management into the 21st Century*, provided some helpful scientific background and thoughtful insights into relevant technical issues and it has made the database it developed available to other workers in the field. It generated special maps, demonstrations of data applications, guidelines for identifying high priority habitat sites, methods for identifying alternative levee locations and new understandings of the influence of important variables.[7]

The SAST Report discusses the available data bases and proposes improvements in their collection, expansion, coordination and accessibility. It describes and discusses the hydrology and physiology of the upper Mississippi River basin and the geomorphology and ecology of its floodplains, and it recommends some steps to take to improve our understanding in these areas. It takes a tentative look at the ways we record and examine economic damages and the impact of upland management and levee protection on flooding, raising important points about both our data and our methodologies. Finally, it makes important recommendations on ways that we can improve our data bases and our analytical tools. Those recommendations include the following:

- Establishing an interagency clearinghouse for data (physical, ecological, social and economic, as well as data on critical infrastructure) useful to management of the basin under a team of government scientists;
- Studying the impact of flooding on the mobilization of hazardous substances;
- Federal state and local agency cooperation in development of more accurate hydrologic and hydraulic models and analytical procedures;

- Completion of the National Wildlife Inventory;
- Identification of the ecologic impacts of the 1993 flood;
- Further study on aspects of the stage-discharge relationship as affected by temperature, channel modifications, etc.;
- Extensive ecologic research including species community and habitat inventories, development of a basin-wide ecologic model and establishment of experimental ecologic restoration sites;
- Developing new models to evaluate the flood reduction impacts of upland measures;
- Performing an inventory and analysis of levee failures;
- Establishing reliable roughness coefficients for natural, urban and suburban floodplain conditions;
- Improving the reliability of the UNET model by providing additional and better data.

THE UNIFIED NATIONAL PROGRAM

The first Unified National Program was published in 1966 as House Document 465, *A Unified National Program for Managing Flood Losses*. In 1968, Section 1302c of the National Flood Insurance Act mandated that the President transmit to Congress "any further proposals necessary for... a unified program, including proposals for the allocation of costs among beneficiaries of flood protection." Subsequent Unified National Programs were prepared in 1976, 1979 and 1986. The most recent *Unified National Program for Floodplain Management 1994* was submitted to Congress in March 1995. The body responsible for developing the policy, the Federal Interagency Floodplain Management Task Force (FIFMTF) is currently chaired by and housed in FEMA, but it was first established under the direction of the Water Resources Council in 1975 and moved to FEMA when the Council was abolished.

As the title implies, the main intent of the 1966 document was to reduce flood losses. It made 17 specific recommendations (under 16 headings) for steps to take. The 1976, 1979 and 1986 documents recorded the progress that had been made in implementing each of them. These recommendations were organized under five objectives:

- To improve basic knowledge about flood hazard;
- To coordinate and plan new developments on the floodplain;
- To provide technical services to managers of floodplain property;
- To establish a national program for flood insurance; and
- To adjust federal flood control policy to sound criteria and changing needs.

The 1976 Unified National Program said that six of the recommendations had been largely implemented, that there had been some progress on eight, but that little had been accomplished on three.[8] No progress was recorded as having been accomplished between 1976 and 1979 but by 1986 seven had been largely implemented, some progress had been made on nine and nothing had been accomplished on only one. The failure in the group comes as no surprise—the recommendation to create a national program to collect flood damage data. The 1994 report does not address the earlier recommendations, as such, nor does it identify flood damage data collection as a continuing problem.

The 1976 revisions changed the name of the program from "managing flood losses" to "floodplain management" and pointed out that new initiatives had been accomplished: a flood insurance program had been established; flood disaster preparedness planning had been authorized by the Disaster Relief Act of 1974; water treatment facility planning grants had been made available in 1972 by PL 92-500; Section 404 of the Clean Water Act now required dredge and fill permits; Coastal Zone Management had been initiated; cost sharing had been extended in principle to nonstructural flood control measures: the Principles and Standards had been published and the National Environmental Policy Act had been passed. The problems in floodplain management highlighted in the 1976 document were fragmented responsibility in floodplain management, over-reliance on public investment and the inability to resolve conflicts of private property rights with state and national interests.[9] The document proposed a three-pronged strategy that would not only modify flooding by building structures, but also modify susceptibility to flood damage by nonstructural measures such as floodplain regulations, flood proof and forecasting, and modify the impact of flooding by helping people prepare, survive and recover. The 1976 document, which was never submitted to Congress, also recommended a conceptual framework for coordinating agencies in the federal and state governments.

The 1979 Unified National Program, prepared at the end of the environmental decade, said our policy included two new goals: restoration of an environment in which natural functions can again operate and prevention of alteration to natural functions of floodplains. It recognized the "natural and beneficial values" of floodplains and it said that the best means of preserving and protecting remaining natural values is to avoid development within floodplains. It updated the 1976 document by including the issuance of executive orders on floodplain management (EO 11988) and wetlands protection (EO 11990) and other presidential actions and it was submitted to Congress.

The 1986 statement of the Unified National Program pointed to new federal emphasis on mitigation activities and made 22 recommendations for future action, half of which were directed to state and local governments for their implementation.[10] The document made proposals to states, for example, for legislation, executive orders, agency organization and program expansions. It proposed some operational changes to local governments along with the recommendation that they adopt and enforce zoning, subdivision and building codes to manage floodplains.

The 1994 unified program states clearly that floodplains are managed for two purposes: (1) to reduce the loss of life, disruption and damages caused by floods; and (2) to preserve and restore the natural resources of the floodplain. It records a long list of strategies and tools for floodplain management, organized around the four traditional strategies of modifying flooding, modifying susceptibility to flood damage, modifying the impact of flooding and protecting the natural resources. The 1994 document points out that "with the addition of the goals, timetable and evaluation mechanisms introduced in this revision, the Unified National Program for Floodplain Management supplies a much-needed sense of national accomplishment and direction."[11] Four goals are identified:

- Formalize a national goal setting and monitoring system;
- Reduce, by at least half, the risks to life, property and the natural resources of the nation's floodplain;
- Develop and implement a process to encourage positive attitudes toward floodplain management;
- Establish in-house floodplain management capability nationwide.

Fifteen actions are identified to specifically implement these goals,[12] with a completion date assigned to each action. The evaluation mechanisms are to be developed, presumably, in the three actions associated with the "goal-setting and monitoring" goal, which is to be accomplished by 1997, along with the four actions associated with the training goal. The four agenda items to encourage positive attitudes are to be completed by 1997. The most action-oriented goal, to reduce actual risks, specifies that both the economic damages for our highest risk floodplain structures and the potential for degradation of natural resources would at least be cut in half by 2020.

CURRENT FLOODPLAIN MANAGEMENT POLICY

Our floodplain management goals today are to reduce and prevent the economic damages caused by floods and to preserve and restore the natural ecosystems of our floodplains. Reduction of property damage has always been a national goal, one that we have so far failed to achieve. Loss of life, however, has been dramatically reduced below the levels of 90 years ago, no doubt because of our vastly improved warning and communication systems. To the extent that national economic development and improved employment opportunities were once part of our explicitly stated floodplain management goals, they no longer are. Ecologic preservation is a more recent goal, emphatically explicit in the 1994 Unified National Program.

The managers of our floodplains, today, are the local, state and federal government, each playing a particular role and within the federal government, a number of different agencies coordinated by the FIFMTF. We have come a long way since the beginning of the 19th century, when private property owners generally took matters into their own hands, organizing into drainage districts to build levees and drain floodplains. Gradually the federal government assumed control, first of navigational activities and later of specific flood control works on individual river stretches where severe flood damage and loss of life had been experienced. By 1936 the federal government had accepted responsibility for the prevention of flood damages on all of our nation's rivers. Land use regulation, however, is outside the federal jurisdiction, since local governments and individual property owners control the uses to which the floodplain is put, and so federal policy today stresses the importance of the state and local roles. The federal government attempts to influence local land use decisions. The state assumes a variety of roles, particularly in multi-objective and river basin planning.

Flood control was nothing more, for many years, than dams and levees to keep the water off the floodplain, but today our floodplain management strategies include a wide array of tools. In the 1960s nonstructural measures became popular, and it was envisioned that floodplain regulations could prevent new people from moving onto the floodplain, that flood proofing, flood warning and insurance would help the flood victims who were already there and that disaster assistance would be reserved for those few who could not be otherwise helped. The most recent tool in the nonstructural category has been mitigation: paying part of the cost to remove the flood victim from the floodplain or stop the farmer from growing crops. Although most floodplain management strategies are designed to accomplish the economic goal there are a few acquisition programs designed to benefit wildlife habitat. While mitigation programs that remove buildings and cropland from the floodplain may provide indirect environmental benefits, the removal of human activity does not automatically restore the floodplain to its natural condition. The 1994 Unified National Program begins, at least, by calling for inventories of all natural resources in both metropolitan and rural areas. These are the goals we want to reach, the ways we want to reach them and the managers we expect to make it happen. We have a long way to go.

REFERENCES

1. Interagency Floodplain Management Review Committee. Sharing the Challenge: Floodplain Management into the 21st Century. Report to the Administration Floodplain Management Task Force, 1994, xii.
2. *Ibid*, 66.
3. *Ibid*, 65.
4. *Ibid*, 82.
5. *Ibid*, 113.
6. *Ibid*, 115.
7. Interagency Floodplain Management Review Committee. Science for Floodplain Management into the 21st Century. Preliminary Report of the Scientific Assessment and Strategy Team to the Administration Floodplain Management Task Force, 1994:5.
8. LR Johnston Associates. Floodplain Management in the United States: An Assessment Report, Volume 2. Federal Interagency Floodplain Management Task Force, 1992:5-5.
9. *Ibid*, 5-4,5,6.
10. *Ibid*, 5-13.
11. Federal Interagency Floodplain Management Task Force. A Unified National Program for Floodplain Management, Washington, DC 1994:24.
12. *Ibid*, 31-33.

CHAPTER 8

The Practice of Floodplain Management

The national investments in solving our flood-related problems have been substantial. Our understanding of the causes of flooding has broadened, and our ideas about their solutions have expanded. Since the 1960s we have had a Unified National Program for Floodplain Management, updated recently, which lays out an impressive array of national goals and strategies. Our actual performance, however, has not kept pace with our ideas.

One reason that the theory of floodplain management has moved out ahead of its practice is that we have no way of knowing whether anything is working, in terms of economic, let alone ecologic, benefits. We do not measure economic damages, we cannot measure ecologic damages and our procedures for developing the benefit cost (B/C) ratios that underpin our project selection are severely flawed. Our technology for predicting flood damages and designing projects to prevent them is too sophisticated for floodplain residents to understand, yet this technology conceals serious deficiencies and may not serve our purposes well.

Our goals for floodplain management have changed dramatically since we first began to deal with floods in the early 19th century. Where once we were concerned exclusively with loss of lives and property, today we also appreciate the value of our floodplain ecosystems and are committed to protecting them. Yet we have few programs to accomplish this. Our attempts to purchase floodplain lands and thus protect them from commercial development are thwarted by farm program subsidies and aid that artificially inflate floodplain property values.

Since the early 1960s we have understood that the structural containment of floods can never keep up with flood damages as long as development continues to escalate on the floodplains, yet floodplain development continues to escalate. The federal and some state governments may have policies supporting floodplain regulation, but they have no authority to implement them. Local governments, which have the authority, have no interest in exercising it. The private owners of floodplain properties have few incentives to abandon their use, pushing away government regulation with one hand, receiving government benefits with the other. The National Flood Insurance Program (NFIP) was a wonderful idea, twenty-eight years ago, and no one is willing today to face the implications of the fact that it has not accomplished what it was designed to do. And, finally, we recognize the risk of providing too much aid to flood victims, in that it encourages them to remain on the floodplain, yet every time it floods politics overrules sound program thinking and aid is distributed more generously than before.

In 1989 a National Review Committee prepared an *Action Agenda for Managing the Nation's Floodplains* in response to FEMA's *Status Report on the Nation's Floodplain Management Activity. A Status Report on the Nation's Floodplain Management Activities: An Interim Report* was made available for review in April 1989, as a preliminary document to the 1992 L.R. Johnston and Associates Assessment Report. The Committee was formed by the Federal Interagency Floodplain Management Task Force, composed of eleven outside experts, chaired by Gilbert F. White and co-chaired by William E. Riebsame. The "Situation In Brief" described in the report said that "Despite massive public and private efforts to reduce flood vulnerability, losses to the nation from occupancy of riverine and coastal areas subject to inundation are continuing to escalate in constant dollars... Most can be attributed to increased property at risk. Vulnerable property clearly is expanding in extent and value...The [damage] statistics are notoriously incomplete and inaccurate...continuing flood damages and losses stem from the ways floodplains are used... The current system for managing floodplains and protecting the nation from impacts of unwise use is piecemeal."[1] It commented a few pages later in the report that "The Unified National Program is neither unified nor national."[2]

The Action Agenda made some specific program assessments, identified a number of factors affecting further activity and proposed six sets of actions:

- Integrating floodplain management strategies into comprehensive planning;
- Improving the data base;
- Giving weight to local conditions;
- Minimizing federal program conflicts;
- Reducing vulnerability of existing buildings; and
- Providing better training of floodplain managers.

The report described the problems with unusual candor, but the actions that it proposed to solve them do not appear to have sufficient force to match the magnitude of those problems. Seven years have passed—there's been a major flood, an extensive floodplain management assessment report and a new Unified National Program. Other than that, things have not changed.

TECHNICAL PROBLEMS

It is hard to believe that we still have no reliable system for collecting economic damages caused by flood events. Gilbert White has been asking for good damage estimates since the 1940s. In 1986, on the occasion of the 50th anniversary of the 1936 Flood Control Act, White pointed out that even though a committee was established in 1939 under the National Resources Committee to develop a system of collecting uniform damage data, and such a proposal was made again in 1966, no such system exists today. He concludes that "there is good reason to suggest that damages generally are overestimated."[3] Without good damage data we cannot describe our flooding problems, we cannot evaluate the programs designed to solve them and, in fact, we will never know when and where they may have actually been solved. We rely, instead, upon media reports, anecdotal evidence and the estimates of experts in the field, without knowing what is being described or why. Even our estimates of lives lost are probably inaccurate. In the 1993 midwestern floods, for example, the official estimate of loss of life is 47, but a Missouri report[4] records 31 for that state alone, only one of the nine affected.

After years of tolerating inadequate and perhaps even misleading flood damage measurements, the U.S. Army Corps of Engineers (Corps) made a commitment to do a comprehensive damage assessment after the 1993 floods. If it appears at all, and there is no indication that it will, the assessment will come out at least three years after the event, long after other, erroneous numbers have been entered into the official record and burned into the public mind. Perhaps the resources are not there to complete the work, perhaps the job is much more difficult than expected, perhaps the interest in doing the job is low or, perhaps, all three. In early 1996 the project manager was on assignment in Korea.

A comprehensive measurement of ecologic damages has never been done in association with a major flood event and probably never will be. The first step would be to do the baseline surveys on the 154 million flood-prone acres identified in the 1992 National Resources Inventory (see chapter 2). It would be difficult enough to find the resources to do the initial job, much less to keep it updated. The other half of the process would be to reinventory after the flood event. Although the environmental resources inventory done by the Corps in their 1995 Floodplain Management Assessment is chock full of interesting data on soils and species, land cover and land use, aquatic resources and fisheries, wildlife and wildlife management areas, natural and recreation areas, the one category that is empty is *effects of flooding*. "While many studies are underway," it explains, "few have been completed at this time."[5] If we don't have the capability to collect the economic damages, it's unlikely that we will ever be able to collect the ecologic ones.

The *Principles and Standards* (P&S) issued in 1983 declared that federal water resources projects must consider the impact on the natural resources of the floodplain, but they were superseded ten years later by the *Principles and Guidelines* (P&G), which limited our water resources planning to the single objective of contributing to national economic development (NED). Since NED must be calculated in monetary units, it is not possible to perform genuine ecologic assessments in the context of the P&G. In 1986 Bory Steinberg surveyed the over 170 flood control new starts since NEPA was enacted in 1969 and found that only three included major nonstructural components.[6] Steinberg said the two reasons the Corps gave for not implementing nonstructural measures were local resistance and economic unfeasibility. Since many of the benefits of nonstructural measures are ecologic, and ecologic cannot be factored into the B/C ratio that determines economic feasibility, this outcome is not surprising.

Without good economic damage data we can never know how well our flood damage prevention projects and programs work or how accurate our B/C ratios are in describing them. Benefit/cost ratios are, by their nature, imprecise. The measured costs are incomplete. As we have seen above, they never include ecologic damages in the calculation. They even fail as economic indicators, because they do not recognize and measure the extent to which a project produces accelerating residual damages. The drama of increasing residual damages has been understood for years, yet it continues. It is acted out in the context of each new flood control project as, today, in Sacramento, California there are plans for major new development in the Natomas Basin, as soon as 100-year protection is achieved. Extensive development is planned for the Basin.[7] The cost side of the equation is clearly inadequate.

The benefit (economic damages prevented) side of the equation is subject, as well, to numerous uncertainties and imperfections. In chapter 6 the process involved in generating B/C ratios is described. At each step of the way the possibility for major error exists. The relationship between discharge and frequency is based upon limited historical records and is often not updated. On the upper Mississippi River, for example, where many new peak discharges were established in the 1993 floods, there had been no update on the discharge-frequency curve since 1979.[8]

The relationship between the stage and the flow is generated out of approximations of data representing channel conveyance, a measurement derived from consideration of river cross sections, geometric calculations and "roughness" that may not always be appropriate. The FPMA study concluded, for example, that if agricultural levees in the middle Mississippi and the lower Missouri River were to be removed, reductions in flood stages would be largely influenced by the use put to the floodplain and that agricultural uses would lower stages (from -3 to +1 feet) more reliably than would forested floodplains (-3 to +4.5 feet). These results were very sensitive to the value selected to represent the channel conveyance or roughness, a constant known as Manning's "n" value. The FPMA analysis gave values of 0.08 to agricultural uses of the floodplain and 0.32 to forested cover for the Missouri River stretch examined in the study. A report published by an interagency federal scientific team convened after the 1993 floods, the SAST Report, discusses this procedure and concludes that "for the no-agricultural-levee analysis the determination of the proper hydraulic roughness coefficient or Manning's "n" value became critically important."[9] The report goes on to point out that the highest value for forested cover used by the USGS was only 0.20. The values used in the FPMA study were actually those recommended by SAST as a compromise, but the report goes on to recommend that research should be done to determine the proper "n" values for a variety of land covers, including urban and suburban, and to look at changes in roughness depending on the size of the flood: water flowing through tree trunks, for example, would move more rapidly than when the upper level was among the branches and leaves. Temperature also has a big effect on the relationship between flow and stage, and the stage of a river resulting from a given flow varies from season to season. Temperature effect has been shown, for example, to produce a stage difference of six feet for a 400,000 cubic feet per second flowing past St. Louis (SAST).[10] Sometimes, for unknown reasons, the relationship between flow and stage may change over time. In St. Joseph, Missouri, for example, flows of 100,000 cubic feet per second never exceeded elevations of 17 feet, between 1928 and 1959. Since 1959, however, those same sized flows have exceeded 17 feet in height on 16 different occasions.[11]

The HEC-2 model traditionally used by many agencies to compute flood stages, is a steady state model that is considered inadequate in the analysis of nonstructural floodplain management techniques. Williams points out, for example, that its bias "offers little incentive to protect or restore floodplains."[12] It is being used far less than in the past, and the recent studies discussed in these pages have largely replaced it with the more appropriate UNET model, as pointed out in chapter 6. Modeling techniques and analytical procedures have important impacts on and implications for floodplain management decision making.

Delineation of the floodplain is based, in turn, on these hydrologic calculations. Applying the estimates of river stages to a given floodplain encounters other major problems; topographical mapping is inaccurate or represented by approximates and rarely updated even as changes occur in the floodplain profile and its uses.

Economic assumptions are made about damageable properties within the floodplain defined in this manner that may themselves be incorrect or out of date. Estimates of property use and value are calculated by formula and do not reflect individual variations or changes over time in floodplain use. Economic damages such as crop losses, for example, do not recognize the characteristics of the actual floodplain but rather, are calculated according to formulas. Neither does a single damage figure take into consideration the characteristics of a particular flood such as its energy, wave action, sediment loads or duration of inundation, all of which will affect damages. The Corps suggests that damage calculations for the 1993 flood underestimated the actual damages, which were incurred during long periods of inundation.[13]

Even if the initial calculation of the damages is reliable, conditions may be different at the time of a flood event. In Tucson, Arizona, for example, a flood-induced scouring of the river bottom allowed a 53,000 cubic feet per second (cfs) flood through a channel whose previous capacity had been only 40,000 cfs, while the opposite effect occurred below Phoenix on the Gila River, where another flood deposited ten feet of sediment in the main channel and sent what had been considered to be a 100-year flood out to inundate the 500-year floodplain.[14]

Reservoirs and other flood control structures may not always be operated in accordance with the original design. Upstream from Sacramento on the American River, prior to the 1986 floods, the operating decisions at the Folsom Dam may have reduced the 120-year design protection downstream by as much as 50%. The operators apparently failed to make the necessary releases in the early stages of the event because not only were they unwilling to inundate the downstream recreational facilities, but also they were not convinced a flood of the predicted magnitude would actually occur.[15] Likewise in 1993 city workers failed to implement a strategic element in the Des Moines flood protection plan by neglecting to sandbag a railroad right-of-way, allowing 58 blocks of downtown Des Moines to flood. Although the procedure was on the books, it had never been necessary to execute before.[16]

The Corps of Engineers has recently developed a methodology for dealing with problems of "risk and uncertainty" (Corps of Engineers, EC 1105-2-205, 1994) that is discussed at length in the National Research Council (NRC) report on the American River since the methodology was apparently used on that project in what the NRC believes to be the first such major application.[17] The NRC report describes the methodology and its concepts at some length and at a level too technical for interpretation here. Some important points, however, are worth repeating.

The distinction between risk and uncertainty is, generally, the difference between values that we do not know because they can only be described in terms of a probability and those that have specific values that we can't or don't want to ascertain. Risk describes the random variability in a natural process such as precipitation; uncertainty describes data that could be difficult to collect, such as the exact number of homes on a floodplain.

The Corps describes their new procedure, according to the NRC report to be "... similar to present practice but differs in that uncertainty is explicitly quantified and integrated into the analysis."[18] The report reviewed the procedure and concluded that 'the proposed Corps of Engineers risk and uncertainty methodology... in the calculation of average flood risk and the average annual flood damages... inflates those estimates. This upward bias is a concern if the methodology is adopted nationwide because it could distort the economic evaluation of projects."[19] The Review Committee that prepared the NRC report found the Corps' application of the methodology to the American River study to be "particularly confusing"[20] and was not able to determine how it was used in the work they were examining. The American River project and the NRC report are discussed in greater detail in the next chapter.

SOCIO-PSYCHOLOGICAL PROBLEMS

The way people who are at risk of flooding think about flooding has always been considered a roadblock to good floodplain management. The Action Agenda says that the fact that "people believe it won't happen again, or in their community" is a constraint to public action to improve floodplain management.[21] Typically, public information and education are proposed as the solution, but attitudes may be more resistant to information about long-term and low probability risk than we have optimistically assumed.

Howard Kunreuther has described a "natural disaster syndrome" that applies to such events as floods and hurricanes.[22] He points out that people don't use protective devices or buy voluntary insurance because they underestimate the probability of occurrence and believe "it won't happen to me." Below a certain probability level, apparently, people tend to behave as if the risk were zero, in dealing with all kinds of protective devices. Further, Kunreuther says, there is an unwillingness to pay up-front in anticipation of future benefits or to pay more than is perceived as "too high." Kunreuther conducted research on a group of people who were no more willing, for example, to spend money on a dead bolt lock if they had a 5-year lease than if their lease was for only one year. He sees as representative of this same behavior the fact that at the time of the 1993 Midwest flood only 42,000, or 5% of the 803,000 flood-prone households in the region had purchased flood insurance policies,[23] and that one in every five NFIP policy holders cancels his coverage each year. In the context of these studies it seems highly improbable that people would spend much money in anticipation of an event they understand to have a 1 in 100 probability of occurring and since the cost of insurance for properties with the highest risk of flooding is actuarially based, it's not surprising that property owners pass it up.

It is generally believed that disaster assistance programs create what the insurance industry calls a *moral hazard,* a condition that stimulates risky behavior because it is perceived to remove the consequences of the risk. When low-cost subsidized flood insurance and generous disaster assistance is available to floodplain dwellers who suffer damages, common sense would tell us that they become less motivated to move or stay out of flood prone areas. Interestingly, Kunreuther has so far found no empirical evidence for this conventional wisdom: "...the data we do have suggests that most homeowners, when asked how they expect to finance recovery, expect to rely on their own resources or bank loans."[24]

The way people think about flooding and its causes help to mold their ideas of appropriate solutions. One way that people think is that flooding is someone else's fault. Government policies are blamed, or upper watershed and upstream practices. Flooded homeowners on the lower Illinois River have claimed, for example, that they knew when the floodwater in their basin was from Chicago, a result of opening up the upstream locks at Joliet (personal experience). Proponents of nonstructural measures blame Corps construction projects for flooding; people who farm on floodplains call for more reservoirs in the upper watershed and better management of the ones already there. Attempts to garner attitudes toward flooding, which normally are surveyed after floods among those who suffered most, elicit irrational responses that are probably closer to the mark than the carefully constructed and moderate statements made by organizations and agencies. The following are a few examples of public reaction to the Corps study after the 1993 floods on the Mississippi:[25]

> *"You're affecting the lives of a lot of people yet your ideas are so slanted toward fish and wildlife it makes me wonder who has priority, fish or people!"*

> *"We don't want to be made into a wetland. We don't need so much studying, what we need is our levee repaired and not so many agencies telling us how to take care of our lives."*

> *"I don't think that we can afford to think of abandoning our levee system and all its benefits to our people and our economic system."*

"Wetlands are a waste of government money which will come from hard working citizens anyway."

"I think [the levees] are very unsightly, they should make them better looking and we'll never mind them. They cost too much money to build. They're a waste of time. If someone wants to live in a flood zone screw them. It's their fault they moved next to them not mine. So move the dumb things."

"I would like to see the levees gone. Let the rivers run their natural courses, without being bottled up. I as a taxpayer am tired of buying people out time and again because they are too stupid to realize that the river is at their back doors when they buy. It amazes me to think that you people need our comments when you have engineers being paid to tell you the levees need to go... Use your heads instead of my wallet!"

"The river flood plains should be used for the river—let it flood when it wants to. Man has always tried to down rate nature, at great sacrifice and costs (to the taxpayers). If you need any help in tearing down the levees, please call me."

"This is a very good example of a communist state. Great numbers of the population who derive their living from the river bottoms are against your proposal but a few people in power think they know what is better."

Whatever drives behavior in the floodplain, it derives more from the gut than from the head and it would be wise not to lay too much stock in measures dependent upon rational behavior of flood victims. Peoples' homes and work, in other words their lives, are affected when it floods. We can expect our engineers and policy decision makers, perhaps, to understand and respond to technical niceties such as probabilities of risk, but it's unreasonable to expect the same from actual or potential flood victims.

POLITICAL PROBLEMS

Floodplain management only has a high priority as a public problem when it floods. Action initiated in times of crisis, such as the Galloway Report and 1994 Unified National Program, sink rapidly into obscurity after the floodplains dry out. No action has been taken on the 1994 Unified Program, although the 1995 deadline for one action, development of a mechanism for implementation, has come and gone, and there are no plans to implement the second action, a national forum, whose completion date is 1996. As it turns out, although the Unified National Program for 1994 was actually submitted to Congress, it is not expected to be acted on by the Administration and will be supplemented by a different document to be issued at an unspecified future date (personal communication with Acting Chairman of the Federal Interagency Floodplain Management Task Force, January 17, 1996). Decisions on that program, whatever it may be, will no doubt be postponed until after crises of greater import are disposed. In spite of all the policy reviews and proposals that were done after the 1993 floods, the present Administration had no officially adopted program in February, 1996.

In times of flood crisis, however, even the officially adopted programs are swept aside by congressional enthusiasms to comfort their constituents. Flood victims exert exceptional pressures on the political system. The publicity associated with flooding associates an innocence and helplessness with their suffering which prompts legislators to override program and policy limits designed to penalize those among them who previously made irresponsible personal decisions. This only ensures that such decisions and other equally bad ones will be made again. After the 1993 flood it would not have been surprising, considering the congressional exceptions made to certain programs, if people had increased their damageable property on the floodplain and not bought flood insurance; if drainage districts hadn't bothered to repair and maintain levees to Corps standards, and if local governments had refused to participate in cost sharing. These irresponsible behaviors were all rewarded by a congress pandering for votes.

JURISDICTIONAL PROBLEMS

Most recent policy statements have pointed to a confusion of governmental roles, not least the Galloway Report, which listed this confusion as one of the three major problems facing floodplain management. It's not so much confusion as contrariness. The federal government, who pays for the damages, wants local governments to keep development off the floodplains. The local governments and the private property owners who are their constituents want protection so they can continue to develop the floodplain. The basis for the local government resistance to floodplain regulation is the private property issue, which has come to the foreground, recently, in 104th Congressional deliberations on the "takings" issue. Imposing a restriction on the use a person can make of his property is being considered equivalent to a economic 'taking' of that property and thus deserving of compensation. In their 1995 survey of state floodplain managers, the Association of State Floodplain Managers asked if local officials were more reluctant today to adopt and enforce floodplain management regulations for fear they might be challenged as "takings." Only six of the respondents said "not reluctant at all," 17 said "somewhat" and 15 (almost twice as many as in the 1992 survey) answered "very reluctant."[26] A Lou Harris nationwide poll with a 3% margin of error, conducted in April 1995, asked if the federal government should have the right to set regulations to bar development and use of private lands, and 59% answered "no." Yet when the same respondents were asked the question again, this time indicating that such use would harm the environment, 79% said "yes."[27]

The Galloway Report quotes a representative of state floodplain managers as calling for a change in the federal floodplain management role. What the states want is that the federal agencies share planning and funding with the local and state levels of government but even were this done, it would not address the critical issue. For as long as the federal government has been calling for local governments to restrict new development on the floodplains, with or without the NFIP, local governments have been permitting it. In a study of two national samples conducted in

1978 and 1982, Burby concluded that "Local floodplain regulations... have had little effect on the location of new development within or outside of flood hazard areas."[28] In growing cities such as Sacramento California, for example, floodplains may provide the only remaining opportunities for centrally located industrial, commercial and residential development. Shoreline floodplains become prime locations for valuable residential and vacation condominiums. Rural floodplains provide rich soils and flat undisturbed areas for crop production. The floodplain is exactly where people want to be. Property owners are unwilling to sacrifice the economic opportunities and they resist regulatory and zoning practices that would limit new development. Gilbert White provides a wonderful description of local floodplain regulation in Boulder, Colorado, in his address to the symposium to commemorate the 50th anniversary of the 1936 Flood Control Act:[29]

> *Consider the community of Boulder, Colorado, which has its own set of regulations that, among other provisions, prohibit new residential development in the floodways of Boulder Creek as well as hazardous tributaries, and require new residences within the 100-year flood plain to be two feet above the estimated one percent flood. Long prior to that enactment it had built its municipal building midway between the creek channel and the levee line proposed by the Corps of Engineers for protection under the 1936 act. Subsequently the city erected a new public library in the floodway, is contemplating a "flood-proofed" extension to the library, and has permitted extensive nonresidential encroachment in the floodplains. The city does have a well-designed and as yet untested flood warning system and is a member of the Denver Urban Drainage and Flood Control District with somewhat similar regulations.*
>
> *The University of Colorado as a state agency operates independently of municipal and district regulations. It has built part of a married-student housing development in a sector of the flood plain, is developing or acquiring high-density residential housing in the flood plain using a two-foot elevation from the Federal Emergency Management Agency standards, is building a new house for its president at an elevation two feet above a one percent flood in a tributary floodway and is*

talking about a research campus extending across the floodway. Such actions are not fully in compliance with city and district guidelines and in the evaluation of alternatives are believed by some to be contrary to the intent of a 1988 executive order by the governor of the state."

Local governments alone have the authority to regulate land use but, since they represent landowners, they can't be expected to do it. Decision makers are unwilling not only to risk the political liability of angering floodplain property owners, but also to deprive their municipality of the additional property taxes from expanded economic and residential development. Finally, local resources are too limited to fund floodplain management at the level of competence and efficacy necessary to overcome the obstacles. Steinberg claimed that one of the two reasons that the Corps had failed to implement nonstructural solutions (see above) was that local people didn't want them. That's a rather important reason.

There is reason to believe, moreover, that every level of government behaves as they did in Boulder, Colorado. The very high costs of public damages in floods cannot be all attributable to roads and other unmovable elements of infrastructure. It never occurred to anyone in one town totally submerged by the Mississippi River floods in 1993, not to rebuild the United States Post Office exactly where it had been before.

INSTITUTIONAL PROBLEMS

By any count, and there are several, the shift of financial responsibility from the property owner to the federal government by making subsidized flood insurance available isn't working. In 1987 Burby and Kaiser estimated that between five and seven million structures were "at risk" and two million policies had been sold in 1980. They suggested that the reason sales did not increase significantly in the next four years was because the NFIP had gradually increased the premiums, shifting costs from the public to the private sector, a major goal of the program.[30] In 1993 Kunreuther reported an estimate that 2 million out of 9.6 flood prone households had policies. The NFIP itself has announced the goal of increasing policies in force by 20% between October 1994 and September 1996 and reports that policies in October 1994 were just above 2.8 million and had reached 3 million by the spring of 1995, even though it also estimates that this number represents no more than 25% of structures located in high risk flood areas (personal communication). It has been a struggle for the NFIP to accomplish these increases in coverage, and the proportion of people reached is still very low. Increasing the number of policies, however, will not solve two other problems: First, that 41% of all policies are still subsidized and second, that almost half (44%) of all payments made by the NFIP since 1978 have been to buildings damaged at least twice.[31]

One of the reasons there has been no progress in nonstructural floodplain management, it has been argued, is that it is being managed by engineers. This argument follows the lines of "if your tool is a hammer, all problems are nails," and the tool of the Corps is construction. Yet the agency has been asked to do a variety of difficult and ill-defined tasks, the most recent of which being the FPMA. After the 1993 midwestern floods, for example, half a dozen senators stepped forward and in the first few minutes of testimony received by the Committee on Environment and Public Works of the U.S. Senate called upon the Corps representative that was present to accomplish a number of things, including the following:[32]

- "Develop a new strategy, one that will take our economy and our natural resources into the next century..."
- "Protect people from floods and keep barge and port traffic flowing..."
- "Safeguard the rivers and the fish in them, the wetlands that surround them..."
- "As manager of lakes that last year attracted more than 200 million visitors, the Corps must invest in recreation..."
- "Look closely at their environmental and recreation missions..."
- "Approach the Corps' missions from the broader context of watershed management..."
- "Take a comprehensive look at the upper Mississippi..."
- "Look at the constitutionally protected property rights of the people who, acting in good faith through many

> years and with the express statutory direction of the Corps, have begun farming operations, bought farmland, improved it and provided very valuable agricultural crops from which they've paid Federal taxes, State taxes and supported their local governments..."

The Corps has responded, as best it could, to the changing and often conflicting mandates of Congress. It is expected to incorporate ecologic benefits into its B/C ratio calculations, without a mechanism for doing so. It is expected to implement watershed planning without the necessary structure or resources. It is expected to elicit local opinion and then it is urged to ignore what it says. Bory Steinberg's comment (above) bears repeating: the Corps has incorporated major nonstructural components into only three projects, because local interests object to them and they can't be justified economically. Under the circumstances, how can we expect the Corps to do otherwise than to continue to build structures?

ECONOMIC PROBLEMS

The only floodplain programs which provide genuine environmental benefits, either by accident or by design, are the buyout programs. They can be prohibitively costly, though, simply because other government programs support and inflate floodplain land values. Farm subsidies, heavily subsidized floodplain insurance programs and flood protection structures all enhance the value of floodplain property. The combination of crop insurance and disaster assistance has been shown, in some cases, to exceed a farmer's expected return during normal conditions.[33] Returned to its natural state the floodplain would have high ecologic and low economic values—the federal government would save money by both avoiding the costly artificial economic supports and paying far less to buy out the properties. Reducing disaster payments would easily produce substantial shifts in floodplain usage, from crops to natural conditions, by prompting farmers to pull their marginal lands out of production. The crop insurance "reforms" are not likely to accomplish this, however, until a premium is charged for coverage.

What our floodplain management programs and policies need is a reality check. Until we collect the data that can accurately evaluate these programs, our talk will be free to venture as far from our deeds as we care to let it. We talk as if we had a set of floodplain management programs that reduce the economic damages and increase the ecologic benefits on our floodplains, but we don't. Some realities, such as our inability to reduce floodplain development, are staring us in the face, but we act in the belief that tomorrow, things will be different. No one collects the data, such as reliable counts of economic damages, changes in floodplain occupancy and federal expenditures on disaster assistance, that would help us see exactly where we are. Without it we can never measure the effectiveness of programs or the progress toward our goals. It's almost as if we didn't want to know.

REFERENCES

1. LR Johnston Associates. Floodplain Management in the United States: An Assessment Report, Volume 2. Federal Interagency Floodplain Management Task Force, 1992: F-5.
2. *Ibid*, F-8.
3. White GF. When may a post-audit teach lessons? In: Rosen H, Reuss M, eds. The Flood Control Challenge: Past, Present and Future. Proceedings of a National Symposium, New Orleans, Louisiana, September 26, 1986. Chicago: Public Works Historical Society, 1988:58.
4. Governor's Task Force on Floodplain Management. Report and Recommendations, July 1994.
5. United States Army Corps of Engineers. Floodplain Management Assessment of the Upper Mississippi River and Lower Missouri Rivers and Tributaries, Appendix C, 1995:1-7.
6. Steinberg B. Flood control in urban areas: past, present and future. In: Rosen H, Reuss M, eds. *op cit,* 97.
7. National Research Council. Flood Risk Management and the American River Basin; An Evaluation., Washington DC: National Academy Press, 1995:166.
8. Interagency Floodplain Management Review Committee. Sharing the Challenge: Floodplain Management into the 21st Century. Report to the Administration Floodplain Management Task Force, 1994:157-158.
9. Interagency Floodplain Management Review Committee, Science for Floodplain Management into the 21st Century. Preliminary Report of the Scientific Assessment and Strategy Team to the Administration Floodplain Management Task Force, 1994:178.

10. *Ibid*, 177.
11. United States General Accounting Office. Midwest Flood: Information on the Performance, Effects, and Control of Levees. Washington, DC: 1995:50.
12. Williams PB. Flood control vs. flood management. Civil Engineering May 1994:53.
13. US Army Corps of Engineers, *op cit,* 3-12.
14. Johnston, *op cit,* 1-10.
15. National Research Council, *op cit,* 46-47.
16. United States General Accounting office, *op cit,* 74.
17. National Research Council, *op cit,* Ch 4.
18. *Ibid*, 137.
19. *Ibid*, 161.
20. *Ibid*, 156.
21. Johnston, *op cit,* F-12.
22. Kureuther H. The Role of Insurance and Regulations in Reducing Losses from Hurricanes and Other Natural Hazards. Philadelphia: Wharton Center for Risk Management and Decision Processes, University of Pennsylvania, 1994.
23. *Ibid*, 4.
24. *Ibid*, 13.
25. US Army Corps of Engineers, *op cit,* Appendix D.
26. Association of State Floodplain Managers. Floodplain Management 1995: State and Local Programs. Madison, Wisconsin: Association of State Floodplain Managers (to be published in 1996).
27. Environmental opinion polls. River Crossings, 1995.
28. Burby, RJ, Kaiser EJ. An Assessment of Urban Floodplain Management in the United States: The Case for Land Acquisition in Comprehensive Floodplain Management. Madison, Wisconsin: Association of State Flood Plain Managers Technical Report #1, 1987:5.
29. White GF. When may a post-audit teach lessons? In: Rosen H, Reuss M, eds. *op cit,* 59.
30. Burby, RJ Kaiser EJ. *op cit,* 5.
31. Interagency Floodplain Management Review Committee. Sharing the Challenge: Floodplain Management into the 21st Century. *op cit,* 124.
32. Hearings before the Committee on Environment and Public Works, United States Senate, 103rd Congress Second Session, May 26, 1994. Washington: US Government Printing Office, 1994.
33. Philippi NS. Spending Federal Flood Control Dollars: Three Case Studies of the 1993 Mississippi River Floods. Chicago, Illinois: Wetlands Research, Inc. 1995.

CHAPTER 9

THE MISSISSIPPI AND AMERICAN RIVERS

The first United States legislation explicitly for flood control was passed in 1917 to reduce flood damages on the Mississippi and Sacramento Rivers. It authorized investments of $45 million on the Mississippi and $5.6 million on the Sacramento River in California to build levees and clear debris from the channels.[1] Congress believed it was accomplishing a "long range and...comprehensive program of flood control" with this important and precedent-setting piece of legislation.[2]

Almost eight decades and countless floods later, the flooding problems of these two basins are still receiving intense investigation. It was the lower Mississippi that required attention in 1917 and it was the upper river recently. It was the Sacramento River that flooded in the early 1900s and it is the American River, the tributary that joins it in downtown Sacramento, that is the subject of controversy today.

Two very different documents to address these problem areas were produced in 1995. *Flood Risk Management and the American River Basin* was prepared by the Committee on Flood Control Alternatives in the American River Basin (Committee) and the *Floodplain Management Assessment of the Upper Mississippi River and Lower Missouri River and Tributaries* (FPMA) was conducted by the U.S. Army Corps of Engineers (Corps). Both reports were mandated by Congress, the former in response to controversy surrounding 1991 Corps proposals for the American River basin, the latter in response to the 1993 floods on the upper Mississippi. The following discussion relates directly to those two documents.

THE AMERICAN RIVER

The Committee was established by the National Research Council in response to a congressional directive to comment on certain issues related to the controversy surrounding Corps proposals for flood control improvements in the American River basin. It was composed of experts, mostly working in the private sector, from across the country. Their work began in October 1993, was completed in the spring of 1995, and was published later that year. The focus of their review was the 1991 American River Watershed Investigation (ARWI) developed by the Corps to present and evaluate alternative solutions to high potential for flooding along the lower river (see Fig. 9.1).

Heavy storms in February 1986 had produced flood flows that almost overtopped the levees that protect more than one quarter of Sacramento's 1.4 million population, its downtown business district and state capitol buildings. The Corps' 1991 proposal for a structural solution, one of a number discussed in the ARWI, had met with heavy criticism, and the agency was already in the process of doing a new evaluation as the Committee began its work.

The stated purpose of the flood control projects evaluated in the ARWI was to protect the $37 billion in damageable property whose "risk" in 1986 had been too close for comfort. The American River flows out of the Sierra Nevada Mountains and across the Sacramento Valley in a southwesterly direction, to join the mainstem Sacramento River in the heart of the city. The downtown properties are protected by levees along the lower river and by flood storage upstream behind the Folsom Dam, a major recreational impoundment with 75 miles of shoreline. The ARWI reviewed the feasibility of 13 flood control measures and ultimately proposed six combinations of four of these measures: the construction of a new (Auburn) flood control dam; expansion of the capacity in the existing (Folsom) dam, increasing the outlet efficiency at Folsom and increasing the channel capacity downstream from Folsom. The first choice of the Corps was an Auburn dry dam with storage large enough to provide 400-year protection to the

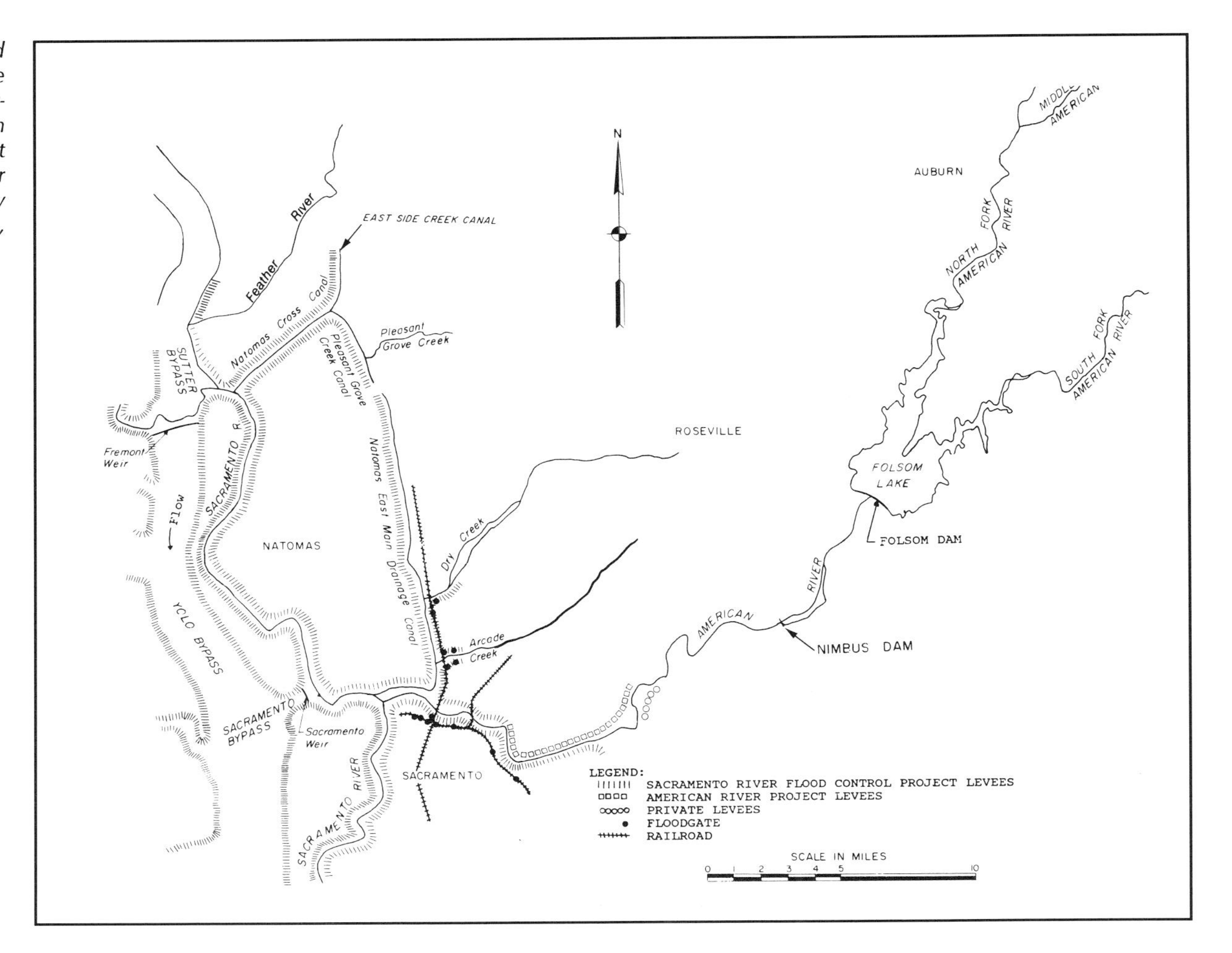

Fig. 9.1. Existing flood control features of the American River Watershed. Reproduced from Flood Risk Management and the American River Basin. National Academy Press, Washington D.C., 1995.

city. The local sponsor, however, preferred a cheaper smaller version of the Auburn dam that would reduce the protection to 200-year. A new 1994 Alternatives Report, which the Committee was able to review as a work-in-progress, offered seven combinations of the same four measures.

One of the earliest in its long history of floods, in 1850, prompted the city to consider moving off the floodplain. It declined to do so and flooding continued (see chapter 2), even after flood control improvements, for the next century. The Folsom Reservoir was designed in 1949, to control what was then understood to be a 500-year storm. The six largest flows between 1907 and 1990, however, had not yet occurred, and the high water flows of 1955, 1963, 1964 and 1986 so changed the frequency-discharge relationship that today the protection that was once considered to be 500-year is closer to the 70-year level. The Folsom Reservoir is not actually managed for flood control. It is upstream from a lot of the rainfall, and has other purposes such as power and recreation. The downstream risk experienced in 1986 was due in part to this multipurpose operational strategy (see the discussion in chapter 8). A dam upstream from Folsom, at Auburn, had been authorized in 1965 and construction actually began, but a 1967 earthquake 45 miles to the north halted the work. A fault was discovered and although the dam was redesigned to reduce seismic risk, the project by then had lost support. Environmental opposition had been mounting and the local sponsor was discouraged by cost sharing formulas in the Water Resources Development Act of 1986 that increased the local share of the costs of the project. The ARWI proposed operating the Auburn reservoir as a dry dam, hoping that it would be viewed as an environmentally more attractive alternative than the proposed 1967 structure. Environmentalists, however, were part of the coalition that prevented the dry dam alternative from being adopted for implementation.

The Committee was specifically asked to address certain technical issues such as the structural stability of the downstream levees. The Corps had concluded that they were safe under the original design flow conditions, yet one consultant argued that they probably were not, and a second concluded that further information was needed to make the determination. The Committee recommended that further study address this issue as well as the issue of channel stability on the lower river. The Committee also proposed additional refinements in hydraulic modeling, including evaluation of a possible dam break at Auburn due to earthquakes. It acknowledged that the Corps' change from the use of the HEC 2 model in 1991 to the UNET in 1994 was an improvement. It disagreed, however, with the new risk and uncertainty analysis procedure used by the Corps in 1994 to calculate the level of protection, described in chapter 8.

The ARWI was criticized for its inadequate environmental impact analysis. The most important unresolved environmental issues surrounded the dry dam, and for this reason the Committee recommended "that a dry dam be used only as a last resort,"[3] and that if it was, gates be installed to prevent rapid drawdowns. Although the dry dam was considered to be an improvement, in terms of potential for ecologic damage, to an Auburn multipurpose reservoir, the Committee concluded that there was not enough information about the potential impacts of periodic inundations behind a dry dam on the plant communities, canyon soils and geologic stability. It saw no way to evaluate the long-term and widespread impacts on relevant ecosystems and pointed out that there was no large dry dam with similar ecologies against which to make comparisons. It indicated that the question of canyon slope stability was a big concern, particularly under the conditions of rapid drawdowns that would be created by an ungated facility, and recommended further study of slope stability. The Committee pointed out that the ARWI used precise estimates of the impact on plant communities where there was no basis for such precision; that some changes such as those in the trout fishery were predicted without any evaluation of their importance and their secondary effect elsewhere; that the regional significance of plant and animal communities hadn't been determined; and that ecologic systems shifts were not explored. The ARWI addressed the environmental problems that had been identified by developing a mitigation plan to compensate for them. The Committee pointed out, however, that mitigation is a poor substitute for integrating environmental interests into the planning process itself, where the damages can be genuinely minimized rather than simply compensated for. The Committee indicated that in at least one case, lack of information made it impossible for the Fish and Wildlife Service to even prepare what they considered to be an appropriate mitigation plan.

In the context of their review of the risk and uncertainty analysis the Committee asked if ecological risk could be assessed as well. It concluded that systems of ecological risk assessment have not been developed sufficiently to be useful and that there are so many gaps in the ecological information available on the American River that "there is little likelihood that such an analysis would be accepted by the scientific and lay community."[4]

A portion of the lower river has been designated under the National Wild and Scenic River System as a sport fishery and a critical recreational resource. There is heavy recreational use of the land adjacent to the river for 23 miles upstream from the city. Lake Folsom is used extensively for fishing, power boating, sail boating and windsurfing, and the Folsom State Recreation Area is one of the most popular parks in the state park system. The upper river has a steep gradient, free flowing white water and relatively natural plant communities. Class IV and V rapids provide exceptional white water rafting opportunities. All of these activities would be affected by various aspects of the flood control measures under consideration and they have competed with one another and with other interests in the quest for flood control solutions.

One of the subjects that the Committee addresses at length is the extent to which any improvement in flood protection will stimulate new economic development and, therefore, increase the risk of residual damages. Although construction of a multipurpose reservoir at Auburn would increase development in the upper watershed, the main concern is a leveed area near downtown Sacramento, the Natomas Basin. The Basin is a 55,000 acre section of the 100-year floodplain, ringed by 41 miles of levees protecting it from the American River and other waterways. A small portion of the southern basin is already developed and other interests are focused on developing the 75 square mile North Natomas region, which the city of Sacramento's general plan identifies as the city's "major growth area for new housing and commercial development."[5] The ground elevation is low and the basin would "fill like a bathtub"[6] according to the Committee, in the event of levees being overtopped or breached.

The developed area of the basin was not subject to National Flood Insurance Program (NFIP) regulations prior to the 1986 flood, because it was considered to have 100-year protection. Subsequently, when the protection level was downgraded to between 40 and 70 years, Congress exempted the area from having to comply with NFIP requirements for another four years. The Committee quotes the NFIP administrator as responding in a personal communication on January 3, 1989 to this precedent-breaking congressional action by pointing out that "maintaining the status quo... will create a significant subsidy for new construction."[7] In response to certain conditions imposed by Congress, however, the City did put a temporary prohibition on new commercial construction in 1990. Although it specified that building permits would not be issued until 100-year protection was reestablished, it permitted economic development planning and construction loan applications for the Natomas Basin. The Committee concludes that the city "is poised to approve the development of North Natomas... despite the unresolved, and perhaps unresolvable, issue of flood hazard."[8]

The local sponsor for the American River project is the Sacramento Area Flood Control Agency (SAFCA). Its strong preference for 200-year protection overruled the Corps' alternative selection, but other public interests prevented even this local selection from getting congressional approval. SAFCA had not represented the upper river canyon protection interests, which opposed any kind of a dam at the Auburn site. Nor had it represented the water supply interests that preferred a full pool multipurpose Auburn reservoir. These two local interest groups, each of which supported very different alternative solutions, united to defeat the official "local" preference. The controversy took on an upstream/downstream flavor, as well, when development interests in Sacramento were accused of destroying the upper river canyons' ecologies, creating what the Committee has called an "untenable deadlock between upper American River public concerns and lower American River public concerns."[9]

The ARWI identified the maximum storage, 400-year protection Auburn reservoir as the alternative that met the National Economic Development (NED) accounting requirements under the *Principles and Guidelines* (P&G), discussed in chapter 4, which meant it had the highest benefit/cost ratio. The Committee does not discuss the details of these benefits, except to indicate that they were entirely flood control and even included a small amount of future benefits in the Natomas Basin. The P&G requirement that an NED plan comply with environmental regulations was met by including the environmental mitigation costs in the calculation of total project costs. The plan eventually recommended by the Corps, however, was the one preferred by the SAFCA that provided 200-year protection and one-fifth fewer net benefits. The P&G allow the NED plan to be supplanted if "...there are overriding reasons for recommending another plan,"[10] and so it was.

Implicit in the Committee's discussion of economics is the suggestion that the American River project has so few national benefits that all the costs should be borne by the recipients of the benefits. The Committee comments generally that the three policy mechanisms by which the federal government forces local governments to accept responsibility for flood risk are requirements that they share in project costs, restrict future development and pay actuarial rates for NFIP insurance premiums, and it concludes that "such balancing is the essence of flood risk management."[11] But according to the Committee's own definition, no such balance existed in the SAFCA case: "In the committee's view, the decision of SAFCA to choose a reduced-size flood risk reduction project, as well as decisions to develop Natomas, has been made in the absence of a requirement to bear reasonable responsibility for the remaining flood risk."[12] It goes a step further when it adds that "benefits for any American River project are not widespread and that SAFCA, by national standards, has a significant ability to pay."[13] The Committee calculates that if the costs were borne entirely by the beneficiaries, property taxes on a $300,000 home would increase by no more than $400 per year for 15 years. The Committee's final recommendations include:

"Before authorizing additional federal financial commitments for flood control on the American River, Congress should explicitly determine whether flood control on the American River constitutes a problem warranting federal involvement based on the presence of widespread national benefits or the limited ability of the community to provide for its own flood protection."[14]

An important part of the document is the general proposal for improvements in the planning process. The Committee makes the point that by failing to include nonflood control interests in its planning process, those interests successfully defeated the selected plan. It points out that institutional alternatives for solving outstanding problems, such as those of water supply and environmental protection, could be as important as strictly engineering alternatives. It concludes with the observation that the major decisions are political and institutional rather than technical, that the question posed is not so much "which alternative?" as it is "what level of risk?" and that such a question must be answered in a nontechnical "willingness to pay" context.

The search for flood protection on the American River as described in the Committee's evaluation document, includes some of the phenomena that were discussed in chapter 8:

- Only structural flood protection measures were considered;
- Levels of protection projections were unreliable: they had dropped significantly over time and even in the current studies, the methodologies had changed and the results were not comparable;
- Project implementation is expected to result in major new development behind the levees;
- The environmental assessment was inadequate and ecologic damages remained unevaluated;
- Congress overruled program requirements that limited risk;
- Upstream environmental interests battled downstream economic interests;
- While local benefits were demonstrable, federal ones were not;

- Technical risk and uncertainty were improperly accounted for in evaluating project alternatives;
- A methodology has not yet been established for assessing ecologic risk.

THE UPPER MISSISSIPPI RIVER

At about the same time, an entirely different floodplain management document was being generated in five of the Corps of Engineers district offices in the midwest. The Corps embarked upon an assessment of floodplain management practices as they affect the upper Mississippi River basin, in early 1994. In response to heavy flooding on the river during the previous summer, the House Committee on Public Works and Transportation authorized the work by adopting House Resolution #2423 on November 3, 1993. The final FPMA document and its five appendices were published in June, 1995.

The scope of the project was extremely ambitious. First, it promised to evaluate seven nonstructural floodplain management programs and to determine the impact that variations in those programs would have had on the flooding in 1993. It then proposed to do hydraulic modeling to calculate the effect that variations in existing levees and reservoirs would have had on the flooding. It also modeled the different impacts that would have been produced by increased upland management practices or reduced flood fighting.

The FPMA study took an extremely important first step: it identified the 125 flooded counties adjacent to rivers and confined its study to them, although 525 counties had been included in the entire disaster declaration and 475 of those were in the two basins under consideration (Fig. 9.2). By identifying the actual overbank flood damages to provide a baseline against which to compare projected program changes, the FPMA provides us with the only data on that flood that is appropriate to floodplain management considerations. As was discussed in chapter 3, those damages are very badly misrepresented elsewhere, and the detailed damage report that was supposed to accompany the FPMA has never been completed.

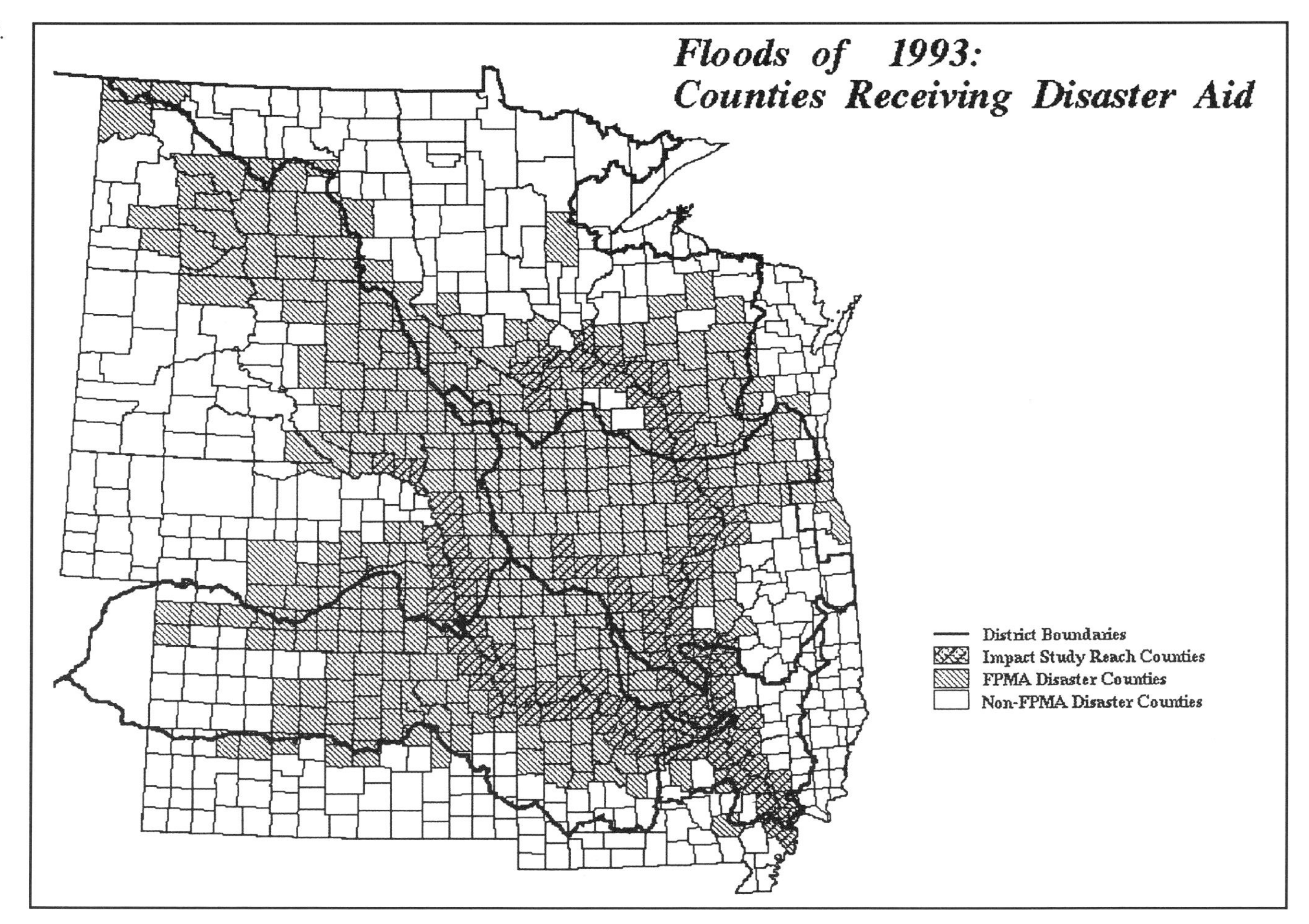
Floods of 1993:
Counties Receiving Disaster Aid
District Boundaries
Impact Study Reach Counties
FPMA Disaster Counties
Non-FPMA Disaster Counties

Fig. 9.2.

The impact of the flood on the 125 counties was described quantitatively as:

- $3,087,394 of economic damages;
- $1,583,262 of federal costs;
- 2,685,281 acres of inundated floodplain;
- 365,285 acres of nonforested wetlands;
- 534,705 acres of forested wetlands;
- 392,512 acres of public lands and 485 recreation sites inundated;
- 1,415 critical facilities at risk;
- 134,849 people directly affected;
- 293 communities flooded; and
- 42,743 residential structures damaged or at severe risk.

These are the "baseline conditions," although it is not always clear whether the definitions of these categories are consistent among all five of the participating Corps districts, each of which collected, displayed and discussed its results separately in the report.

The nonstructural programs that were evaluated are the NFIP, state programs, local programs, federal mitigation programs, disaster relief, wetland restoration and agricultural programs. The report hypothesized three sets of changes (scenarios) in each of the programs and reported how these changes affected the baseline conditions displaying them in matrices (Figs. 9.3, 9.4 and 9.5). This section is extremely disappointing. The hypothesized changes in the programs are uneven in appropriateness, any quantification of impacts was impossible in most cases and the entire process proceeds very unsystematically. The findings are consistent with opinions generally held by workers in the field, and aren't tied to any research results. Among them are the following conclusions on how program changes would have affected the effects of the 1993 floods:

- An expansion of the NFIP would have reduced the need for disaster assistance;
- Local floodplain management measures would have had no significant impact;
- Protecting critical facilities with 500-year protection would have eliminated damages in this category; in-

creased state funding of mitigation programs would have reduced overall damages;

- Increasing the federal share of mitigation costs from 50 to 75% would have had an impact;
- Shifting disaster relief costs to other governments and private entities would have reduced costs;
- Wetland restoration programs that increased wetland restoration by 10 to 25% would have high environmental benefits and low impact (8%) on total floodplain cropland; and
- Shifting from agricultural disaster assistance to crop insurance may not have reduced government expenditures because the insurance is subsidized.

The kind of rigorous analysis of program components that is really needed, unfortunately, is far too complex to have been done within the limited time, fiscal and institutional constraints of this assessment.

The hydraulic modeling produced more interesting results. Four variations on agricultural levees were tested throughout the study area: (1) removing all levees; (2) setting all levees back to open up the floodway to at least 5000 feet; (3) sizing all levees for 25-year protection by raising the lower ones and notching the higher ones; and (4) raising all levees to the height that would have actually confined the 1993 flood. Analyses were also done for two other hypothetical conditions: the removal of all reservoirs and the reduction of runoff from the upper watershed 5 and 10%, to simulate land management and wetland restoration measures. Again, the results are reported separately for each of the five Corps districts, and it is sometimes difficult to tie the final conclusions to the individual analyses.

The FPMA concluded overall that the alternatives of limiting flood fighting, removing levees and downsizing them to 25-year levees would "appear to have little net potential for reducing flood impacts."[15] Flood stages went down and with them urban damage, while agricultural damages and environmental benefits went up, with dramatic increases in total area flooded. Limiting agricultural levees to a 25-year level of protection would have produced a three and a half foot drop in stages on the middle Mississippi and a

SCENARIO CATEGORIES

	IMPACT CATEGORIES	A Base Cond. [All Disast. Counties]	B Base Cond. [FPMA Imp. Counties]	C National Flood Ins. Program Regs.	D State Fldpln. Mgm & Zoning	E Local Fldpln. Mg & Zoning	F Relocation, Mitigation Programs	G Disaster Relief Programs	H Floodplain Wetland Restor. Pro	I Agricultural Support Policies	J Signif. Findings
	ECONOMIC (1,000 $'s)		[1]								
	Flood Damages										
1	Residential (Urban)	$760,892	$662,008	LOW	– LOW	– LOW	–200,000	–< 5%	0	0	
2	Other (Urban)	$1,612,543	$1,447,322	LOW	– LOW	– LOW	–15,000	–< 5%	0	0	
3	Agricultural	$3,852,701	$817,054	0	0	0	0	0	0	(–) SMALL	
4	Other Rural	$233,648	$161,010	LOW	– LOW	– LOW	0	0	0	NEGLIG.	
	Chg. in Govt.Expend.										
5	Emergen.Resp.Costs	$227,405	$200,663	LOW	– LOW	– LOW	–20%	–< 5%	0	(–) NEGL.	
6	Disaster Relief (Agric.)	$1,160,632	$285,180	0	0	0	0	0	0	0	
7	Disaster Relief (Human R.)	$1,297,474	$551,862	+82,000	– LOW	– LOW	–100,000	–< 5%	0	(–) NEGL.	
8	Flood Insurance (NFIP)	$371,969	$276,496	–82,000	– LOW	– LOW	–10%	–< 5%	0	(–) NEGL.	
9	Flood Insurance (FCIC)	$748,095	$269,061	0	0	0	0	0	0	$64,484	
	Chg.Value of FP Resources										
10	Net Ag RE Values	–	–	0	0	0	0	0	0		
11	Net Urban RE Values	–	–	0	0	0	< 5%	0	0		
	ENVIRONMENTAL										
	Natur.Resour.(# acres)										
12	Non–Forested Wetl. (acres)	–	365,285	0	0	0	0	0	37,000	32,000	
13	Threat.&Endang. (# / Occ.)	–	(281/1,043)	0	0	0			+	0	
14	Forest (acres)	–	534,705	0	0	0			110,000	95,000	
	Natural Fldpln.Functions										
15	Fldpln.inundated (acres)	–	2,685,281	0	0	0			0		
	Cultural										
16	Archeol Impacts (–5 to +5)	–	–1	–1(+1)	0(0)	0(0)	–1(0)	–1(0)	–1(NA)		
16A	Hist.Sites(–5 to +5)		–1	–1(+1)	0(0)	0(0)	–1(0)	–1(0)	–1(NA)		
	Open Space										
17	Public lands (acres)	–	392,512	0	0	0			26,000	13,000	
18	Recreation sites (#)	–	485	0	0	0			25	10	
	REDUCT.OF RISK										
	Critical Facilities										
19	# Facil. w/harmful releases	–	207	0	– LOW	0	–< 5%	0	0		
20	# other critical facilities	–	1,208	0	– LOW	0	–< 10%	0	0		
	Prot./Avoid. of Harm										
21	# people vulnerable	185,630	134,849	LOW	– LOW	– LOW	–20,000	–< 5%	0		
	Social Well Being										
22	# communities vulnerable	433	293	0	– LOW	– LOW	–100	0	0		
23	# resident.struct.vulnerable	56,339	42,743	LOW	– LOW	– LOW	–6,000	–< 5%	0		
	IMPLEMENT. COSTS										
24	Structural Costs	–	–	0	0	0	+$215,000	+< 5%	–	–	
25	Other Costs	–	–	+ MODERATE	+ LOW	+ LOW	+$90,000	0	$190 mil	$108 mil	

[1] Economic impacts collected only at the county level

Fig. 9.3. Matrix developed to display the damage reduction that would have been achieved in the 1993 floods., had low level changes in federal programs (scenario 1) been in place. From: Floodplain Management Assessment of the Upper Mississippi River and Lower Mississippi Rivers and Tributaries, U.S. Army Corps of Engineers, June 1995.

Fig. 9.4. Matrix developed to display the damage reduction that would have been achieved in the 1993 floods, had mid-level changes in federal programs (scenario 2) been in place. From: Floodplain Management Assessment of the Upper Mississippi River and Lower Mississippi Rivers and Tributaries, U.S. Army Corps of Engineers, June 1995.

SCENARIO CATEGORIES

	IMPACT CATEGORIES	A Base Cond. [All Disast. Counties]	B Base Cond. [FPMA Imp. Counties]	C National Flood Ins. Program Regs.	D State Fldpln. Mgmt. & Zoning	E Local Fldpln. Mgmt. & Zoning	F Relocation, Mitigation Programs	G Disaster Relief Programs	H Floodplain Wetland Restor. Pro	I Agricultural Support Policies	J Signif. Findings
	ECONOMIC (1,000 $'s)		[1]								
	Flood Damages										
1	Residential (Urban)	$760,892	$662,008	LOW	– HIGH	– LOW	– < 5%	– < 5%	0	0	
2	Other (Urban)	$1,612,543	$1,447,322	LOW	– HIGH	– LOW	0	– < 10%	0	0	
3	Agricultural	$3,852,701	$817,054	0	0	0	0	0	< –5%	(–)	
4	Other Rural	$233,648	$161,010	LOW	– LOW	– LOW	0	0	0	(+)	
	Chg. in Govt.Expend.										
5	Emergen.Resp.Costs	$227,405	$200,663	LOW	– HIGH	– LOW	– < 5%	– < 15%	< –5%	(–) NEGL.	
6	Disaster Relief (Agric.)	$1,160,632	$285,180	0	0	0	0	0	< –5%	NA	
7	Disaster Relief (Human R.)	$1,297,474	$551,862	LOW	– HIGH	– LOW	– < 5%	–375,000	0	(–) NEGL.	
8	Flood Insurance (NFIP)	$371,969	$276,496	MODERATE	– HIGH	– LOW	– < 5%	+20%	0	(–) NEGL.	
9	Flood Insurance (FCIC)	$748,095	$269,061	0	0	0	0	0	< –5%	NA	
	Chg.Value of FP Resources										
10	Net Ag RE Values	–	–	0	0	0	0	0	< –5%		
11	Net Urban RE Values	–	–	0	0	0	< 5%	– < 5%	0		
	ENVIRONMENTAL										
	Natur.Resour.(# acres)										
12	Non–Forested Wetl. (acres)	–	365,285	0	0	0	0	0	52,000	77,600	
13	Threat.&Endang. (# / Occ.)	–	(281/1,043)	0	0	0	0	0	+	+	
14	Forest (acres)	–	534,705	0	0	0	0	0	157,000	218,000	
	Natural Fldpln.Functions										
15	Fldpln.inundated (acres)	–	2,685,281	0	0	0	0	0			
	Cultural										
16	Archeol Impacts (–5 to +5)	–	–1	–1(+1)	–1(+1)	0(0)	–1(+1)	–1(+1)	–1(?)	–1(NA)	
16A	Hist.Sites(–5 to +5)	–	–1	–1(–2)	–1(–2)	0(0)	–1(–2)	–1(–2)	–1(?)	–1(NA)	
17	Open Space										
17	Public lands (acres)	–	392,512	0	+	0	0	0	80,000	8,000	
18	Recreation sites (#)	–	485	0	+	0			75	5	
	REDUCT.OF RISK										
	Critical Facilities										
19	# Facil. w/harmful releases	–	207	0	– HIGH	– LOW	0	– < 5%	0		
20	# other critical facilities	–	1,208	0	– MODERATE	– LOW	0	– < 5%	0		
21	Prot./Avoid. of Harm	–									
21	# people vulnerable	185,630	134,849	LOW	– HIGH	– LOW	– < 5%	– < 5%	0		
	Social Well Being										
22	# communities vulnerable	433	293	0	– MODERATE	– LOW	0	– < 10%	0		
23	# resident.struct.vulnerable	56,339	42,743	LOW	– HIGH	– LOW	– < 5%	0	0		
	IMPLEMENT. COSTS										
24	Structural Costs	–	–	0	–	–	+ <$10,00	+$50,000	$8 million	$60 million	
25	Other Costs	–	–	LOW	+ HIGH	+ MODERATE	+$600	+$100	$196 mil.	$153 mil.	

[1] Economic impacts collected only at the county level

Fig. 9.5. Matrix developed to display the damage reduction that would have been achieved in the 1993 floods, had high level changes in federal programs (scenario 3) been in place. From: Floodplain Management Assessment of the Upper Mississippi River and Lower Mississippi Rivers and Tributaries, U.S. Army Corps of Engineers, June 1995.

SCENARIO CATEGORIES

	IMPACT CATEGORIES	A Base Cond. [All Disast. Counties]	B Base Cond. [FPMA Imp. Counties]	C National Flood Ins. Program Regs.	D State Fldpln. Mgmt. & Zoning	E Local Fldpln. Mgt & Zoning	F Relocation, Mitigation Programs	G Disaster Relief Programs	H Floodplain Wetland Restor. Prog	I Agricultural Support Policies	J Signif. Findings
	ECONOMIC (1,000 $'s)		[1]								
	Flood Damages										
1	Residential (Urban)	$760,892	$662,008	LOW	– LOW	0	–< 10%	–< 5%	0	0	
2	Other (Urban)	$1,612,543	$1,447,322	LOW	– LOW	0	–< 10%	–< 5%	0	0	
3	Agricultural	$3,852,701	$817,054	0	0	0	0	0	<–10%	(–)	
4	Other Rural	$233,648	$161,010	LOW	– LOW	0	0	0	0	(–)	
	Chg. in Govt. Expend.										
5	Emergen. Resp. Costs	$227,405	$200,663	LOW	– LOW	– LOW	–< 5%	–< 10%	<–10%	(–) NEGL.	
6	Disaster Relief (Agric.)	$1,160,632	$285,180	LOW	0	0	0	0	<–10%	0	
7	Disaster Relief (Human R.)	$1,297,474	$551,862	LOW	– LOW	0	–< 5%	–25%	0	(–) NEGL.	
8	Flood Insurance (NFIP)	$371,969	$276,496	MODERATE	– LOW	– LOW	–< 5%	+ 100%	0	(–) NEGL.	
9	Flood Insurance (FCIC)	$748,095	$269,061	0	0	0	0	0	<–10%	$64,484	
	Chg. Value of FP Resources										
10	Net Ag RE Values	–	–	0	0	0	0	0	<–10%		
11	Net Urban RE Values	–	–	0	– MODERATE	0	< 5%	–< 5%	0		
	ENVIRONMENTAL										
	Natur. Resour. (# acres)										
12	Non–Forested Wetl. (acres)	–	365,285	0	+	0	0	0	31,000	+0	
13	Threat. & Endang. (# / Occ.)	–	(281/1,043)	0	+	0	0	0	+	+	
14	Forest (acres)	–	534,705	0	+	0	0	0	94,000	+0	
	Natural Fldpln. Functions										
15	Fldpln. inundated (acres)	–	2,685,281	0	0	0	0	0			
	Cultural										
16	Archeol Impacts (–5 to +5)	–	–1	–1(+2)	–1(0)	0(0)	–1(+2)	–1(+2)	–1(?)	–1(NA)	
16A	Hist. Sites (–5 to +5)	–	–1	–1(–3)	+1(0)	0(0)	–1(–2)	–1(–2)	–1(?)	–1(NA)	
	Open Space										
17	Public lands (acres)	–	392,512	0	+	0	+	0	200,000	+0	
18	Recreation sites (#)	–	485	0	+	0	+	0	100	+0	
	REDUCT. OF RISK										
	Critical Facilities										
19	# Facl. w/harmful releases	–	207	0	– LOW	0	0	–< 5%	0		
20	# other critical facilities	–	1,208	0	– LOW	0	0	–< 5%	0		
21	Prot./Avoid. of Harm										
21	# people vulnerable	185,630	134,849	LOW	– LOW	0	< 10%	+ 25%	0		
	Social Well Being										
22	# communities vulnerable	433	293	0	– LOW	0	–20	+> 5%	0		
23	# resident. struct. vulnerable	56,339	42,743	LOW	– LOW	0	–< 10%	+ 25%	0		
	IMPLEMENT. COSTS										
24	Structural Costs	–	–	0	–	–	+$100,000	+$100,000	$50 million	$100,000	
25	Other Costs	–	–	MODERATE	+ LOW	+ LOW	+$25,000	+$500,000	$50 million	$500,000	

[1] Economic impacts collected only at the county level

two foot drop on the Missouri River near its mouth. If, on the other hand, all the agricultural levees had been raised high enough to contain the 1993 flood, all the agricultural damage would have been prevented, damages on many urban properties would have been increased and it would have cost $5.6 billion to construct, in the St. Louis District alone. Setting back the levees would have had mixed results: overtopping would have been prevented in some districts, but downstream constrictions would have occurred and many of the agricultural benefits that originally justified the levees would have been lost. Most of the districts reported that the removal of existing reservoirs would have caused substantial damage on the major tributaries and overtopping of urban levees; at Kansas City overtopping would have increased the damages in that Corps district by 500%. The Rock Island District pointed out, however, that impacts on the main stem Mississippi would have been minor. The conclusions of the four districts that reported on the impact of reducing upland runoff are confusing: they agreed that flood stages would have been reduced, but didn't translate this into damage reductions. The St. Paul District did describe the positive impacts of these measures as $200 and $400 million respectively, for the 5 and 10% reductions in runoff. It then went on to calculate the cost of accomplishing those benefits, in terms of land acquisition upstream, and those costs are represented in the final analysis as $1.25 and $2.5 billion dollars, respectively.

Translating changes in flood stage to changes in economic damages is hard enough, because it depends upon assumptions about land use and other reactive changes. Relating those same changes to environmental damages and benefits is simply impossible. The only impact that the Corps could quantify was acres inundated. The FPMA acknowledges the problems of evaluating environmental impacts but offers no solutions. It is simply not useful to say that removing levees would result in cropland reverting back to natural conditions, and that runoff reduction could be produced by wetland restoration programs and that both of these actions would have environmental benefits. We know these things and they are not appropriate in the context of a report that purports to assess in quantitative terms. The assessment framework, which is hard to interpret and of limited use in measuring economic impacts, is worthless in evaluating ecologic ones. Since the FPMA fails to "identify the effects of the flooding on the environmental resources within the floodplain," which was claimed as the intent of the Environmental Resources Inventory [16] (see the discussion in chapter 8), it cannot possibly provide any prediction of ecologic changes that might occur as a result of the various alternatives under consideration.

The FPMA incorporated an extensive citizen participation component into the study. Three sets of meetings were conducted throughout the region: 12 of them as the study began in June 1994, then in November another 13 meetings and finally in April 1995, as it was coming to a close, 11 more meetings were held. The June meetings, according to the report, indicated that agricultural interest supported more levees, environmental interests supported nonstructural measures, all interests asked for greater intergovernmental cooperation and everyone wanted to understand the 1993 flood better. The public concerns raised at the November meetings, which were attended by 514 people, apparently included just about everything. The one attended by the author included three public members and ten Corps staff people and provided less information, in fact, than had been distributed the previous week by mail. The April public meetings provided the most information on public concerns. The 543 participants in this final group of meetings were asked to rate a set of programs and actions that roughly reflected the ones considered in the final report. The 365 responses to this request were presented and analyzed separately for each district. The numbers of total responses for the districts were as follows: St. Paul 19, Rock Island 145, Kansas City 76, St. Louis 88 and Omaha 37, and although the report displays the responses from each district, according to the participants' main affiliation, only the Rock Island District participation is large enough to warrant the analysis. The three major affiliations represented in the Rock Island group were agricultural, government and homeowner. There were 87 in the agricultural group, but only 14 each in the homeowner and government groups (the other, unaccounted for interest groups making up the difference), too small a number to tell us much. More than 50% of the

87 agricultural respondents, however, rated the programs and actions as follows:[17]

- Very High: relocation and mitigation programs, and actions that would raise agricultural levees, raise urban levees, protect critical facilities and retain water upland;
- Low: actions that would set back agricultural levees; and
- Very Low: wetland restoration policies and actions that would limit flood fighting and remove agriculture levees.

There is nothing particularly surprising about these results, with one exception: the "very high" value placed on relocation and mitigation programs. These are measures usually associated more with environmental than with agricultural constituencies, and it would be interesting to know how these options were presented, discussed and perceived in the Rock Island District meetings, particularly since none of the small numbers of agricultural respondents in other districts agreed with this rating. The results of the citizen participation would have been more useful if the samples could have been aggregated. The amount of statistical analysis and graphical display performed in Appendix D seems highly unwarranted by the paucity of the data. Six bar charts, for example, were developed to display the information generated from 19 responses.[18] This kind of analytical overkill is misleading.

In spite of the shortcomings in the accomplishment of its lofty purposes, the FPMA report is chock full of useful pieces of information. The attention it gives to critical facilities is particularly interesting. Almost half ($1.447 of the total $3 billion) of the total economic damages on the floodplain were urban but not residential. That category contains "Commercial and industrial structures, public buildings, transportation facilities and utilities."[19] The Corps cautions that the actual count of such facilities, 632, is not reliable[20] but it includes facilities such as *Superfund* sites, waste treatment facilities, federal and state bridges, military installations and schools that would presumably be vulnerable to leverage from federal agencies. It makes some interesting suggestions for solving certain problems by executing strategic tradeoffs such as opening up the downstream floodway as an alternative to building higher urban levees and reducing upland runoff as an alternative to major improvements in an existing levee. The benefits of the FPMA may not end with the production of this report to meet a Congressionally mandated deadline. The report itself, however, takes a very important first step toward defining the problem. Consider the two sets of data outlined in Tables 9.1 and 9.2, taken from the FPMA.[21]

Overbank flooding inundated 2.7 million acres. Sixty-eight percent of the damages were sustained on 3% of the flooded land and about 70% of those weren't even residential. The remaining 32% of the economic damages were sustained on the 56% of inundated land that was agricultural. The remaining 41% of the inundated floodplain, which was described as forest, wetland, water and other, presumably sustained no damages. If the detailed damage report is ever published, it should make possible an even bigger step forward in defining what really happened in the summer of 1993.

The FPMA tells us what we can't do, at least without unlimited time and resources. It did ask the right questions, however, and if the answers aren't worth very much, in their present state,

Table 9.1.

Floodplain Damages, 1993 Floods	Millions of Dollars	Percent
Urban residential	$662	21%
Urban other	$1447	47%
Rural/Agricultural	$978	32%
Total	$3087	100%

Table 9.2.

Land Use	Acres Flooded	Percent Flooded
Urban	73,920	3%
Agricultural	1,502,685	56%
Other	1,108,676	41%
Total	2,685,281	100%

they can perhaps be built upon by other agencies or perhaps applied to other areas. The study made it clear that we haven't developed a way to evaluate the economic impacts of our policies and our nonstructural programs and that the way we measure the economic impacts of our protection projects cannot be used to measure the ecologic impacts. Of course we knew that already. What is surprising is that the Corps agreed to do the project at all. It is less surprising that Congress authorized it, and one wonders, what will they make of it?

REFERENCES

1. Reuss, M. Introduction. In: Rosen H, Reuss M, eds. The Flood Control Challenge: Past, Present and Future. Proceedings of a National Symposium, New Orleans, Louisiana, September 26, 1986. Chicago: Public Works Historical Society, 1988: x.
2. Moore DP, Moore JW. The Army Corps of Engineers and the Evolution of Federal Floodplain Management Policy. Boulder: Institute of Behavioral Science, University of Colorado, 1989:14.
3. National Research Council. Flood Risk Management and the American River Basin; An Evaluation, Washington DC: National Academy Press, 1995:102.
4. *Ibid*, 160.
5. *Ibid*, 169.
6. *Ibid*, 170.
7. *Ibid*, 173.
8. *Ibid*, 175.
9. *Ibid*, 103.
10. *Ibid*, 182.
11. *Ibid*, 183.
12. *Ibid*, 185.
13. *Ibid*, 187.
14. *Ibid*, 201.
15. United States Army Corps of Engineers. Floodplain Management Assessment of the Upper Mississippi River and Lower Missouri Rivers and Tributaries, 1995, 9-62.
16. *Ibid*, Appendix C, 1-7.
17. *Ibid*, 11-11-12.
18. *Ibid*, Appendix D, St. Paul District Summary and Statistics.
19. *Ibid*, 3-12.
20. *Ibid*, 3-16.
21. *Ibid*, Chapter 3.

CHAPTER 10

SCENARIOS FOR THE FUTURE

In looking at our floodplain management experience, several troubling observations can be made that, if not true in every respect, must be kept in mind in looking toward the future.

- The more we do, the less we get done. Our massive program of structurally engineered flood control projects has prevented, no doubt, billions of dollars worth of damage. Specific localized flood control problems, in many parts of the country, have been solved. But our national goal has always been to reduce the *total* damages and by all accounts, those damages are growing. The only consolation provided by the fact that damages are growing no faster than is gross national product, is to realize that the situation could be worse. The thrust of the 1936 Flood Control Act and all the legislation that has followed was to reduce the costs to government and the suffering of people who found themselves, through no fault of their own, in danger from floods. Residual flood damages increase after our structural protection projects have been built. We don't have good nonstructural programs to keep development off the floodplain. For a quarter of a century we have said we could reduce damages by reducing development, yet this development continues to occur. For almost as long we've been talking about ecosystem protection, but it's mostly talk. Governments have found no better ways to protect the natural floodplains from disruption and degradation than to buy them up themselves.

- All this is guesswork because we're not sure what is really happening. We don't have data on flood damages so we don't know if, where, when or how they may be increasing or decreasing. Our best guess is that they're increasing in more places or at least by greater numbers than they're decreasing. We don't have data on floodplain uses so we don't know with certainty where and by how much development is increasing, but our best guess is that it's going up. We don't have baseline environmental data for all our floodplain resources and we don't fully understand the ecologic systems involved, so we don't know if we're losing or gaining ground. Our best guess is that we're losing ground. We don't know exactly how much we're spending on disaster assistance, but our best guess is that it is our most expensive floodplain management tool. Our best guesses are probably right, but we can't be sure.
- The goals of our floodplain management programs aren't being met, but since we never audit their performance, we haven't yet admitted that they may not work. We keep on struggling, fine tuning as we go, to make them better. Local governments don't regulate floodplains so, for example, we say they need more training, yet do we have evidence that better training for local officials reduces the amount of new building in the floodplain? People increase the value of their property in the floodplain so we launch information campaigns on the risk of flooding. Have we ever seen flood damages reduced by making information available? The Corps of Engineers keeps building structural projects that increase residual damages and we blame the Corps, but don't the Department of Agriculture's projects have the same effect? We rely on benefit/cost ratios to select our projects and drive our flood control policies but have we ever tested these mathematical constructs to determine that they have any relation to reality?
- Our floodplain management policy moves further and further away from reality. Our claim to manage flood-

plains for ecologic purposes has been in and on the books since the 1970s, but we haven't done it yet. We talk about coordinating ecologic and economic objectives in floodplain management as if they were compatible but there is little evidence that they are. We devise elaborate theoretical constructs for analyzing data when the data is only rudimentary. Nonstructural management measures are far more popular among planners than they are on the floodplain. Federal policy directives define state and local responsibilities, and state and local officials redefine the federal role. The federal government calls for local floodplain regulations, where it has no jurisdiction, and meanwhile rebuilds its own post offices after floodwater subsides. We talk about lofty partnerships with local governments when those local governments can't even find the resources to maintain their responsibilities for stream gaging networks.

THE BENEFIT/COST RATIO

The single question that has preoccupied us most in flood protection policy has always been: how do we select the best projects? Flood protection projects so blatantly improve the economic conditions of individual property owners that it took a century for this country to be willing to accept the fact that flood control could be a matter of the "general welfare." Up until then projects had been built by property owners to protect that property. They paid the costs, received the benefits and made the decisions, in terms of those costs and those benefits, on which projects should be built. Later they joined together into drainage districts and, collectively, made the same decisions in the same way. Their "willingness to pay" was the selective principle.[1]

When the government began to pay the bill and the private property owner continued to benefit, the basis for rational decision making was lost. Water resource projects are notoriously pork barrel. Politicians can and always have rewarded their favorite constituents with flood control and drainage projects if they could get away with it. The 1936 Flood Control Act dealt with this by requiring that the benefit/cost ratio (B/C) of any project must be at least greater than 1.0. It provided no mechanism for determining what the very best project was, but it could draw a line between those projects that were good enough and those that weren't. The question of level of protection was not addressed.

The 1973 Principles and Standards (P&S) took policy a giant step forward. They called for evaluating water projects in the context of not only the National Economic Development (NED) account, which is basically the B/C ratio, but also the Environmental Quality, Social Well Being and Regional Development accounts. The latter three accounts were cumbersome, if not impossible, to apply, and the P&S were replaced, ten years later, by the Principles and Guidelines (P&G). The P&G elevated the importance of the NED account and required that the preferred project not only have a B/C greater than 1.0, but also that it have the *highest* B/C ratio of the proposed alternatives, regardless of total cost, as long as environmental regulations were enforced. The question of level of protection was still not addressed.

In the American River basin the NED project provided 400-year protection, but the costs were so high that the local sponsor opted for an alternative with only 200-year protection. Since the P&G allow local preferences to overrule, the result was the selection of an alternative other than the NED, with lower B/C ratio and lesser benefits. The American River project addressed the environmental damages by including the costs of mitigation for those damages on the cost side of the ratio. The local sponsor, SACFC, was criticized in the National Research Council Report for acting irresponsibly when they selected the lower protection level. The report pointed out that a lower level of protection, particularly when associated with the new development planned in the community, would ultimately result in higher future residual costs. The SACFC, however, in acting to lower the overall project costs, behaved the way cost sharing policy hopes local governments will act: protective of the public purse. If the locals had no financial responsibility, cost sharing policies assume, they will always want the highest level of protection. In the case of the American River cost sharing policies worked as they were intended to, but questions of level of protection and residual damages complicated the issue, as well they should have.

Residual damages may be, in retrospect, the most important costs of any protection project. The way we evaluate a project, long after it has been built, is by looking at the amount of new development that was subsequently built under its protection, and at the size of the damages that were sustained when floods greater than the design level occurred. The implication of the Committee, in criticizing the SAFCA decision for being irresponsible, was that the local sponsor did not take into account the increases in residual damages that would result from a lower level of protection. Since the actual cost of residual damages is disaster assistance, the costs and blame would fall upon the federal government. There is no mechanism in flood control planning for anticipating and addressing residual damages.

The P&S provides a mechanism for the federal government to get the most value for its money, regardless of total costs, by the requirement that the NED project be selected. When local sponsors make other choices in favor of lower costs, this principle may be violated in favor of the principle of keeping costs down. A third principle for selecting projects was described in the American Rivers report, that of imposing an upper limit on the level of protection provided by the federal government. This does not address the question of how to select the best project as much as it does the question of how much responsibility the federal government should assume in any project. Just because a project is the "best" by any definition, does this mean that the federal government should build it, regardless of the protection level it provides?

LEVEL OF PROTECTION

It is generally believed and often stated that the federal government has assumed the responsibility to protect its citizens from the 100-year flood, but this is not the case. It is only true that the National Flood Insurance Program adopted the 100-year flood as the level of danger that it wants local governments to regulate against. The program considers the area within the 100-year floodplain to be sufficiently at risk of flooding that people should be discouraged from living and building there, in the interests of reducing flood damages. Presumably there is a benefit/cost principle operating here, an assumption that the risks within this floodplain are great enough to justify the development of a public program that imposes all kinds of costs on various levels of government and private citizens.

Federally funded flood protection is provided, however, for all levels of risk. The American River alternative with the greatest B/C ratio provided a 400-year level of protection and even the cheaper alternative selected by the local sponsor provided 200-year protection. The argument that an American River project funded by the general public should provide protection no greater than from the 100-year flood was argued by the U.S. Environmental Protection Agency, which followed with the proposal that selection among those alternatives that met the 100-year criterion should be for the alternative that inflicts the least amount of environmental damage.[2] Such a selection procedure would incorporate, for the first time, a genuine mechanism for limiting environmental degradation. Current procedures put no constraints on ecologic damages other than that they must be paid for. Further, limiting the level of flood protection would result in smaller projects that, almost by definition, have less potential for causing environmental damage. The NED principle of maximum cost efficiency would be sacrificed, however, and the issue of residual damages would not be addressed at all.

In sorting through the issues reviewed in the American River report, it becomes apparent that there are no guiding principles governing the expenditure of federal funds in Sacramento that guarantee either a given level of flood protection, a maximum degree of cost efficiency, a minimum of environmental damage, a minimum total cost or a limit to residual damages. For this reason, no doubt, the Review Committee gently poses to Congress the question of why the public should be taxed to increase Sacramento's flood protection at all. It implies that if Sacramento wants the benefits of flood control, they should be willing to pay for it themselves. It's tempting, certainly, to try to shift the fiscal responsibility for flood control to the local beneficiaries, but in doing so, the federal government would still be held liable for residual damages and lose any leverage it might have to provide protection to ecologic systems.

ECOLOGIC CONSIDERATIONS

The B/C ratio does not and cannot reflect true ecologic costs or benefits. Ecologic values will be better understood and described as more sophisticated accounting systems are developed, but they can never be translated into useful numbers for the calculation of a B/C ratio. A ratio may incorporate the costs of mitigating for environmental damages, but mitigation is qualitatively different from protection. Further, it can be said generally that the greater the level of economic protection, the more environmental damage will be sustained. Yet the ecologic values are the only ones in floodplain management that are truly "public" and deserve the attention of the federal government. Even if local beneficiaries were willing to pay for economic benefits, we can be sure that they will never invest a single cent in ecologic benefits.

WHAT WE KNOW

There are a few lessons that we should have learned by now:

- Discharge-frequency relationships aren't very reliable, for many reasons. The American River Review Committee looked for evidence of nonrandomness in flooding to explain the change in discharge-frequency relationships after 1959. If there is no randomness in flooding, describing floods in terms of probabilities and therefore, frequencies would be pointless. Randomness, however, being the characteristic of an infinite number of events, may not be very helpful in predicting floods within the short time frames in which we operate, and the precision with which we describe the floods is misleading. Generally speaking we can feel comfortable that larger floods occur less frequently than smaller floods, but to draw a line on the ground and say that water will reach this point, on average, once in one-hundred years, is nonsense. Whether or not we have a handle on the risk, we can be sure that our information and our technology isn't that good yet.
- Local governments aren't going to regulate growth on the floodplain if they can get away with it, and federal

proclamations that they should won't change the situation. Local government decision making is driven largely by the need for economic development and tax base expansion. Boulder and Sacramento are just two examples of communities whose needs drive them to make floodplain management decisions counter to public policy at the national level.

- People who live and work in floodplains don't worry about it when it's dry and suffer and get angry when it isn't. They do not make rational decisions in terms of risk in relation to flooding anymore than they do in other parts of their lives. If they did, for example, no one would let their child ride a bike, where the odds of a head injury are 1 in 7, and no one would be afraid to fly, where the odds of being killed are 1 in 2.2 million.[3] If someone's home is on the floodplain the operative concept is *home*, not *floodplain*.
- There is no system for translating ecologic into monetary values and probably never will be. The fact that we need such a system doesn't mean that it's possible to construct. We would better spend our time eliminating that need than to keep struggling to accomplish the impossible. As long as we use the B/C ratio as the basis for selecting flood control projects there will be no genuine consideration of ecologic values in the planning process.
- The greatest enemy of good floodplain management is disaster assistance. The only good strategy we have to protect floodplain ecosystems—land acquisition—is thwarted by other government programs that artificially inflate floodplain property values.
- The National Flood Insurance Program has failed to meet its goals, in spite of all the good that it has done. Floodplains have been mapped, communities have regulated development on the floodplains, property owners have shared the cost of living there by buying actuarially based policies; all these things have happened and the understaffed and underfinanced program offices struggle to push the numbers higher. We may be win-

ning some of the battles, but the war is being lost. Floodplain development has not been stopped and the responsibility for flood damage losses has not been shifted off the federal government and on to property owners. After twenty-eight years, if the NFIP could have done these things, it probably would have.

- Politicians will always end run the system. We may have succeeded in removing most of the pork barrel, but the political overrides of program requirements, each time it floods, undermine the best ideas and efforts of public programs. The federal government is never free to ask, in the broadest public interest, what the best program or project might be; individual legislators instead, press for projects that favor their constituencies and trade among themselves for special interest benefits. This trading process may, if all the partners have equal power, lead to an equitable distribution of benefits that, itself, is a kind of public interest. It could even be argued, perhaps, that such a process occurred in the American River valley. The process is so inefficient, however, that it can't keep pace with the floodplain management problems it tries to solve.
- There will be less and less money to spend on floodplain management programs. Total federal expenditures in 1992 were divided, roughly, as follows: (l) 52% for direct cash payments to individuals; (2) 13% for interest on the public debt; (3) 18% for protection of the country from outside attack; (4) 1% for international programs; and (5) 16% for all programs to improve the public welfare. The Congressional Budget Office (CBO) projects that the portion of our taxes spent on the first two categories could increase to 65% and 16%, respectively, by the year 2005, leaving only 19% to fund the remaining three categories.[4] Under these conditions, according to CBO projections, total federal debt would have doubled from 3.4 to 6.8 trillion dollars and the annual deficit would have increased from $203 to $462 billion. These projections will never be realized, hopefully, because deficit reduction is a high priority of both

> Congress and the Administration. Enormous savings will have to be accomplished somewhere, however, and inevitably floodplain management programs will be among the many that feel the impacts.

Bad public policies, like bad marriages, are based on the unhappy expectation that human nature can be changed. It may be in this sense that many aspects of floodplain management policy are bad. We must remind ourselves that politicians, local governments and floodplain home owners, to say nothing of engineers and environmentalists, will continue to be themselves.

SCENARIOS FOR THE FUTURE

Realistically, if external conditions do not change in substantial ways, the most likely scenario for the future is that the business of floodplain management will continue pretty much as it has. Efforts such as the Galloway Report may have small impacts upon existing programs but, while many of these programs are good, they are probably not good enough to help us reach our goals. New residential and commercial development will expand the potential damages in the unprotected areas of the floodplain and errors in our calculations and extreme precipitation events will raise the toll of residual damages behind levees and below reservoirs each time their levels of design protection are exceeded, which will always be more often than we expect. Increases in ecologic damages will probably taper off as fewer new projects are built, but only small incremental gains will be accomplished through floodplain acquisition programs. The thrust of our programs will be, in dollars spent, disaster assistance. In the context of this discouraging scenario, the best we can try to do is to:

- Learn more about the problems we are trying to solve. Collect better data on economic damages and better understanding of ecologic systems. Forget about monetizing ecologic values.
- Use the benefit/cost ratio as a guide rather than a standard. Replace it with a program planning process that gives equal weight to all the institutional interests, particularly ecology.
- Reduce all disaster aid—individual, public and agricultural—by as much as we can get away with.

External conditions may change, substantially, however, providing the possibility of at least two other scenarios. The possibilities of a real budget crisis loom larger with each succeeding failure at deficit reduction in Washington. Dramatic fiscal action by the federal government could go in two directions: one good and one bad for floodplain management.

The worst scenario is that the federal budget will be balanced by making larger cuts in programs than in direct payments (entitlements). Under such a scenario the individual, public and agricultural disaster assistance would continue, inflating floodplain land values and relieving floodplain property owners of all responsibility for the risk of where they live and work. Funds for structural projects, mitigation and acquisition would dry up. The NFIP and other nonstructural programs would be maintained at funding levels too low to be effective. The first programs to go, no doubt, would be the few we have for data collection and analysis, since it would not be an environment where we would want to know what was going on out there. The incentives for this scenario are political: the recipients of government checks are more reliable political supporters than are program advocates.

A better scenario in response to a federal budget crisis is to cut the entitlement and program budgets in direct proportion to their size. Larger programs would be cut the most to achieve the greatest savings. Flood protection structures would no longer be built to avoid the initial costs. Above all, disaster payments would be reduced, motivating private citizens, businesses and farmers to stay off the floodplains, and state and local governments to protect their own property. Floodplain property values would properly reflect the risk of flooding. The incentives for this scenario are purely fiscal: economic damages from project costs, residual damages, floodplain acquisition and disaster assistance would be drastically reduced; and with them, so would ecologic damages.

Ecologic values will never drive private or public program when money is tight. Ecologic goals can be accomplished, however, by programs that recognize the economic realities of floodplain management. Those realities are that: (1) protection projects generate costly residual damages that aren't calculated in B/C ratios, yet without calculating those damages, B/C ratios don't account for all the economic costs; and that (2) disaster assistance programs cost us far more than the amount of the outlays themselves: they artificially inflate floodplain property, make acquisition programs too costly and provide a bias against nonstructural solutions. The economic best use of our floodplains is probably, if we took all these factors into account, to give the rural floodplains back to the natural systems that thrive there and to provide what cash we can spare to helping the people and businesses in greatest need to move to higher ground.

Ecologic interests cannot compete with economic ones on their own grounds. Their case can benefit from more data collection, better understanding of natural ecosystems and a greater role in the planning process. It cannot benefit from efforts to tie into the B/C process, which should be discarded on economic grounds alone. The 100-year regulatory floodplain is too unreliable to be enforced uniformly and our dependence on technical analysis of inadequate data to settle what are more correctly institutional and political disputes has been carried too far. The time is long overdue for us to turn our futile attempt to engineer the floodplain back on itself and, for economic reasons, provide some ecologic benefits.

"For tis the sport to have the engineer
Hoist with his own petar."
(Hamlet, III. iv)

REFERENCES

1. Shabman L. The benefits and costs of flood control: reflections on the flood control act of 1936. In: Rosen H, Reuss M, eds. The Flood Control Challenge: Past, Present and Future. Proceedings of a National Symposium, New Orleans, Louisiana, September 26, 1986. Chicago: Public Works Historical Society, 1988:112.
2. National Research Council. Flood Risk Management and the American River Basin; An Evaluation. Washington DC: National Academy Press, 1995:182.
3. Shook MD, Shook RL. The Book of Odds. New York: Penguin, 1991.
4. Congressional Budget Office. Economic and Budget Outlook, August 1995. Washington DC: 1995:22.

APPENDIX A

List of Acronyms and Abbreviations

ARWI	American River Watershed Investigation (conducted by the Corps, 1991)
B/C	Benefit/Cost Ratio
Corps	United States Army Corps of Engineers (USACE)
EO	Executive Order (issued by the President)
EWRP	Emergency Wetland Reserve Program, a program of the USDA
FEMA	Federal Emergency Management Agency
FIFMTF	Federal Interagency Floodplain Management Task Force
FPMA	Floodplain Management Assessment of the Upper Mississippi River and Lower Missouri Rivers and Tributaries, June 1995, United States Army Corps of Engineers
FWS	Fish and Wildlife Service, within the United States Department of Interior
GNP	Gross National Product
LMV	Lower Mississippi Valley, a Division of the Corps
NEPA	National Environmental Policy Act

NRB	National Resources Board, later called the National Resources Committee
NED	National Economic Development, an "account" established by the P&S
NFIP	National Flood Insurance Program
NRCS	Natural Resources Conservation Service, formerly the SCS
NRI	National Resources Inventory, conducted by the NRCS
NWS	National Weather Service, part of the National Oceanic and Atmospheric Administration within the United States Department of Commerce
OMB	United States Office of Management and Budget
P&G	Principles and Guidelines
P&S	Principles and Standards
SAFCA	Sacramento Area Flood Control Agency
SAST	Scientific Assessment and Strategy Team
SCS	Soil Conservation Service, within the United States Department of Agriculture, recently renamed the NRCS
TVA	Tennessee Valley Authority
USDA	United States Department of Agriculture
USGS	United States Geological Survey, within the United States Department of Interior
WRC	Water Resources Council
WPA	Works Projects Administration, a program of the "New Deal"
WRP	Wetlands Reserve Program, administered by the USDA

APPENDIX B

Declaration of Policy in the Flood Control Act of 1936*

(In the *Laws of the United States Relating To The Improvement of Rivers and Harbors from August 11, 1790 To January 2, 1939*, 3 volumes: Washington, D.C. Government Printing Office, 1940)

Section 1

It is hereby recognized that destructive floods upon the rivers of the United States, upsetting orderly processes and causing loss of life and property, including the erosion of lands, and impairing and obstructing navigation, highways, railroads and other channels of commerce between the States, constitute a menace to national welfare; that it is the sense of Congress that flood control on navigable waters or their tributaries is a proper activity of the Federal Government in cooperation with States, their political subdivisions and localities thereof; that investigations and improvements of rivers and other waterways, including watersheds thereof, for flood-control purposes are in the interest of the general welfare; that the Federal Government should improve or participate in the improvement of navigable waters or their tributaries, including watershed thereof for flood-control purposes if the benefits to whomsoever they may accrue are in excess of the estimated costs, and if the lives and social security of people are otherwise adversely affected.

* In the *Laws of the United States Relating To The Improvement of Rivers and Harbors from August 11, 1970 to January 2, 1939*, 3 volumes: Washington, D.C. Government Printing Office, 1940.

SECTION 2

That, hereafter, Federal investigations and improvements of rivers and other waterways for flood control and allied purposes shall be under the jurisdiction of and shall be prosecuted by the War Department under the direction of the Secretary of War and supervision of the Chief of Engineers, and federal investigations of watersheds and measures for run-off and waterflow retardation and soil erosion prevention on watersheds shall be under the jurisdiction of and shall be prosecuted by the Department of Agriculture under the direction of the Secretary of Agriculture, except as otherwise provided by Act of Congress; and that in their reports upon examinations and surveys, the Secretary of War and the Secretary of Agriculture shall be guided as to flood-control measures by the principles set forth in section 1 in the determination of the Federal interests involved: *Provided,* That the foregoing grants of authority shall not interfere with investigations and river improvements incident to reclamation projects that may now be in progress or may be hereafter undertaken by the Bureau of Reclamation of the Interior Department pursuant to any general or specific authorization of law.

SECTION 3

That hereafter no money appropriated under authority of the Act shall be expended on the construction of any project until States, political subdivisions thereof, or other responsible local agencies have given assurances satisfactory to the Secretary of War that they will (a) provide without cost to the United States all lands, easements, and rights-of-way necessary for the construction of the project, except as otherwise provided herein; (b) hold and save the United States free from damages due to the construction works; (c) maintain and operate all the works after completion in accordance with regulations prescribed by the Secretary of War: *Provided,* That the construction of any dam authorized herein may be undertaken without delay when the dam site has been acquired and the assurances prescribed herein have been furnished without awaiting the acquisition of the easements and rights-of-way required for the reservoir area: *And provided further,* That whenever expenditures for lands, easements and rights-of-way by States, political subdivisions thereof, or responsible local agencies for any individual project or useful part thereof shall have exceeded the present estimated construction cost therefore, the local agency concerned may be reimbursed one-half of its excess expenditures over said estimated construction cost: *And provided further*, That when benefits of any project or useful part thereof accrue to lands and property outside of the State in which said project or part thereof is located, the Secretary of War with the consent of the State wherein the same are located may acquire the necessary lands, easements and rights-of-way for said project or part thereof after he has received from the States, political subdivisions thereof, or responsible local agencies benefited the present estimated cost of said lands, easements and rights-of-way less one-half the amount by which the estimated cost of these lands, easements and rights-of-way exceeds the estimated construction costs corresponding thereto: *And provided further*, That the Secretary of War shall determine the proportion of the present estimated costs of said lands, easements and rights-of-way that each State, political subdivision thereof, or responsible local agency should contribute in consideration for the benefits to be received by such agencies: *And provided further*, That whenever not less than 75 per centum of the benefits as estimated by the Secretary of War of any project or useful part thereof accrue to land and property outside of the State in which said project or part thereof is located provision (c) of this section shall not apply thereto; nothing herein shall impair or abridge the powers now existing in the Department of War with respect to navigable streams: *And provided further*, That nothing herein shall be construed to interfere with the completion of any reservoir or flood control work authorized by the Congress and now under way.

SECTION 4

The consent of Congress is hereby given to any two or more States to enter into compacts or agreements in connection with any project or operation authorized by this Act for flood control or the prevention of damage to life or property by reason of floods upon any stream or streams and their tributaries which lie in two or more such States, for the purpose of providing, in such manner and such proportion as may be agreed upon by such States and approved by the Secretary of War, funds for construction and maintenance, for the payment of damages, and for the purchase of rights-of-way, lands and easements in connection with such project or operation. No such compact or agreement shall become effective without the further consent or ratification of Congress, except a compact or agreement which provides that all money to be expended and performed by the Department of War, with the exception of such reasonable sums as may be reserved by the States entering into the compact or agreement for the purpose of collecting taxes and maintaining the necessary State organizations for carrying out the compact or agreement.

APPENDIX C

Recommendations and Actions Proposed by the Galloway Report*‡

4.1: Reduce the vulnerability of population centers to damages from the standard project flood discharge.

4.2: Reduce the vulnerability of critical infrastructure to damage from the standard project flood discharge.

5.1: Revise the RCRA locational standards and contingency planning regulations for consideration of flood hazards in areas impacted by the Standard Project Flood.

5.2: Increase the state role in all floodplain management activities including, but not limited to, floodfighting, recovery, hazard mitigation, buyout, floodplain regulation, levee permitting, zoning, enforcement and planning.

5.3: Restructure and refine the scope of federal technical services programs and increase funding for the USACE in the areas of Floodplain Management Services and Planning Assistance to the States programs and increase funding for states through the FEMA Community Assistance Program.

* Excerpted from *Sharing the Challenge: Floodplain Management into the 21st Century.* Report to the Administration Floodplain Management Task Force by the Interagency Floodplain Management Committee, 1994.

‡ "Recommendations address problems that the Review Committee believes merit attention; however, the solutions to these problems can be accomplished within agency resources, existing programs, or cooperative efforts." *(Sharing the Challenge: Floodplain Management into the 21st Century, page xxii.)*

5.4: Hold FEMA's existing disaster assistance cost-sharing requirements to no more than 75/25; seek to make other agencies disaster programs' cost-share requirements consistent at 75/25.

5.5: The Administration should seek increased funding for federal agencies to support collaborative planning participation with other federal agencies.

5.6: Promote the use of programmatic NEPA documents in the planning process.

5.7: OMB should issue a directive that requires periodic reevaluation of federal water resources projects to include potential operation and maintenance modifications.

5.8: OMB should use only the benefit-cost ratio for damage reductions to existing development in establishing Administration funding priorities unless a standard project flood level of protection is provided.

6.1: Enhance predisaster planning and training.

6.2: The FEMA should review its policy of issuing revisions to flood insurance maps which remove property from the floodplain based on fill.

6.3: Federal agencies involved in floodplain management should include information regarding floodplain management and past and probable future flood heights and extents in their education and public affairs initiatives.

6.4: State floodplain management officials should encourage local school districts to include natural hazard education in their curricula.

7.1: The Administration should support increased funding for the Refuge Revenue Sharing Act.

8.1: Federal agencies should capitalize on opportunities, within existing authorities and resources, to enhance the environment when reviewing operations or undertaking repairs or improvements to existing flood damage reduction programs.

8.2: The Administration should propose legislation that establishes consistent cost-sharing across agencies for non-federal participation in like activities.

8.3: The USACE should investigate procedures to minimize impacts associated with levee overtops.

8.4: The USACE should coordinate with the SCS to decide on appropriate criteria for evaluating the economics of levee repairs.

8.5: Maintain flexibility in hazard mitigation programs to promote cost-effective and appropriate mitigation techniques.

8.6: Encourage establishment of state-chaired task forces to coordinate buyouts and other hazard mitigation activities.

8.7: Encourage use of CDBG, EDA and other funding to acquire and relocate or take other mitigation actions where consistent with program objectives.

9.1: Integrate federal flood response and recovery under the FEMA.

9.2: Enhance the linkage among response, recovery, and floodplain management.

9.3: Continue to seek federal-state co-leadership of an interagency hazard mitigation team.

9.4: States should actively encourage flood insurance purchase by their citizens.

10.1: Where they do not already do so, states should assume responsibility for regulating levee-related activities such as levee location, alignment, design, construction, upgrade, maintenance, repair and floodfighting.

10.2: The USACE should consider land acquisition as an alternative during planning and design of habitat rehabilitation and enhancement projects under the Upper Mississippi River Environmental Management Program.

11.1: Federal water agencies, in collaboration with state, tribal, and local entities, should review and update, as necessary, discharge-frequency relationships for streamflow gages in the upper Mississippi River Basin to reflect the 1993 flood data. The adequacy of the existing stream gaging network should also be reviewed.

11.2: Federal agencies, coordinated by NWS and USGS, should collaborate on an assessment of the effectiveness of the stream gaging network and flood forecasting during the 1993 Midwest floods.

11.3: The USACE and USGS should investigate and better define relationships between high energy erosion zones, other zones in floodprone areas, and levee failure.

11.4: Federal agencies should conduct research on biotechnical engineering techniques and incorporate them into design manuals.

11.5: OMB should review the current system of funding disaster relief; consideration should be given to encouraging the National Science Foundation to support such a review.

11.6: USDA should evaluate the impact of federal farm programs on agricultural land use decisions in and out of the floodplain.

11.7: FEMA should conduct research on the issue of NFIP market penetration to determine who buys flood insurance and who does not and why.

11.8: The National Science Foundation should consider funding research on the following subjects:

- Full accounting of all public and private benefits and costs of floodplain occupancy and associated floodplain management measures, including both monetary and non-monetary methods of accounting;
- Mapping and regulating areas with movable stream channels and storm drainage overflow and backup;
- Special impacts of floods, including epidemiological and mental health factors; and
- The feasibility and effectiveness of the use of meteorologic data and geomorphic and botanical evidence in conjunction with hydrologic and hydraulic models to estimate flood frequency.

ACTIONS PROPOSED BY THE GALLOWAY REPORT*

5.1: Enact a national Floodplain Management Act to define governmental responsibilities, strengthen federal-state coordination and assure accountability.

5.2: Revitalize the Water Resources Council.

5.3: Reestablish Basin Commissions in a revised form reflecting current needs.

5.4: Issue a new Executive Order to reaffirm the federal government's commitment to floodplain management with an expanded scope.

5.5: OMB should direct all federal agencies to conduct an assessment of the vulnerability of flooding using a scientific sample of federal facilities and those state and local facilities constructed wholly or in part with federal aid.

5.6: Seek revision of Section 1134 of the Water Resources Development Act of 1986 to provide for phase out of federal leases in the floodplain.

5.7: For communities not participating in the NFIP, limit public assistance grants.

5.8: Encourage communities to obtain private affordable insurance for infrastructure as a prerequisite to receiving public assistance.

5.9: Provide loans for the upgrade of infrastructure and other public facilities.

5.10: Establish as the new, co-equal objectives for planning water resources projects under Principles and Guidelines:

- To enhance national economic development by increasing the value of the Nation's output of goods and services and improving national economic efficiency, and

* "Actions may involve resource commitments beyond an agency's baseline posture." *(Sharing the challenge: Floodplain Management into the 21st century, page xxii.)*

- To enhance the quality of the environment by the management, conservation, preservation, creation, restoration, or improvement of the quality of natural and cultural resources and ecological systems.

5.11: Establish an interdisciplinary, interagency review of the P&G by affected agency representatives to address:

- Structural versus non-structural project bias;
- Inclusion of system of accounts or a similar mechanism for displaying impacts;
- Inclusion of collaborative planning in an ecosystems context for major studies; and
- Expansion of the application of the revised P&G to water and land programs, projects, and activities to include:
  - All federally constructed watershed and water and land programs;
  - National parks and recreational areas;
  - Wild, scenic, recreational rivers and wilderness areas;
  - Wetland and estuary projects and coastal zones; and
  - National refuges.

6.1: The Administration should establish an interagency task force, jointly chaired by the USDA and EPA, to formulate a coordinated, comprehensive approach to multiple objective watershed management.

6.2: The DOI, USDA and EPA should coordinate and support federal riverine and riparian area restoration.

6.3: The Administration's legislative proposals for the 1995 Farm Bill should support continuation and expansion of conservation and voluntary acquisition programs focused on critical lands within watersheds. The proposal should support technical and financial assistance for implementation of watershed management, riparian enhancement, wetland restoration and upland treatment measures.

6.4: Promote the NFIP Community Rating System as a means of encouraging communities to develop floodplain management and hazard mitigation plans and incorporate floodplain management concerns into their ongoing community planning and decision making.

6.5: Provide funding for the development of state and community floodplain management and hazard mitigation plans.

6.6: Map all communities with flood hazard areas that are developed or could be developed.

6.7: To improve and accelerate delivery of NFIP map products, the Administration should propose supplementing those funds obtained for floodplain mapping from NFIP policyholders with appropriated funds.

6.8: Utilize technology to improve floodplain mapping.

7.1: The administration should establish a lead agency for coordinating acquisition of title and easements to lands acquired for environmental purposes.

7.2: The Administration should develop emergency implementation procedures to organize federal agencies for environmental land acquisitions.

7.3: The DOI should formalize environmental considerations in multi-agency disaster recovery land restoration activity through a coordinated Memorandum of Agreement.

7.4: Seek legislative authority for flexibility in use of programmed funds in emergency situations.

7.5: The DOI should focus land acquisition efforts on river reaches and areas with significant habitat values or resource impacts.

7.6: Require agencies to co-fund ecosystem management using Operation and Maintenance funds.

7.7: Enact legislation allowing cost-share participation and eligibility requirements under Sections 906 and 1135 of the 1986 WRDA to include federal, state and non-governmental contributions as well as work in-kind.

7.8: Allocate funds for mitigation lands in concert with and at the same pace as project construction.

8.1: Establish the USACE as the principal federal levee construction agency.

8.2: The Administration should reaffirm its support for the USACE criteria under the PL 84-99 levee repair program and send a clear message that future exceptions will not be made.

8.3: Federal and state officials should restrict support of flood fighting to those levees that have been approved for flood fighting by the USACE.

8.4: Establish a task force to develop common procedures for federal buyouts and mitigation programs.

8.5: Provide states the option of receiving FEMA Section 404 Hazard Mitigation Grants as a block grant.

8.6: Provide funds in major disasters where supplemental appropriations are made for buyouts and hazard mitigation, through FEMA's Section 404 Hazard Mitigation Grant Program.

8.7: Establish a programmatic buyout and hazard mitigation program with funding authorities independent of disaster declarations.

8.8: The FEMA should continue to enforce substantial damage requirements, but decide on a definition of substantial damage and stick to that definition.

8.9: The Administration should support insurance coverage for mitigation actions necessary to comply with local floodplain management regulations.

8.10: Develop a program to reduce losses to repetitively damaged insured properties through insurance surcharges, increased deductibles, mitigation insurance, and/or mitigation actions.

9.1: Hold an interagency strategic planning meeting for those Presidentially declared disasters that require a multi-agency recovery effort.

9.2: Increase NFIP market penetration through improved lender compliance with the mandatory purchase requirement.

9.3: Provide for the escrow of flood insurance premiums or payment plans to help make flood insurance affordable.

9.4: Develop improved marketing techniques.

9.5: Reduce the amount of post-disaster support to those who could have bought flood insurance but did not, to that level needed to provide for immediate health, safety and welfare; provide a safety net for low-income flood victims.

9.6: Require actuarial-based flood insurance behind all levees that provide protection less than the standard project flood.

9.7: Increase the 5-day waiting period for flood insurance coverage to at least 15 days.

9.8: The Administration should continue to support reform of Federal Crop Insurance that limits crop disaster assistance payments, increases participation, and makes the program more actuarially sound.

10.1: Establish upper Mississippi and Missouri basin commissions with a charge to coordinate development and maintenance of comprehensive water resources management plans to include, among other purposes, ecosystem management, flood damage reduction and navigation.

10.2: The Administration should expand the mission of the Mississippi River Commission to include the upper Mississippi and Missouri rivers. Further, to recognize ecosystem management as a co-equal federal interest with flood damage reduction and navigation, the Administration should request legislative change to expand commission membership to include the DOI.

10.3: Assign responsibility for development of an Upper Mississippi River and Tributaries (UMR&T) system plan and for a major maintenance and major rehabilitation program for federally-related levees to an expanded Mississippi River Commission, operating under the USACE.

10.4: Seek approval from the Congress to redirect the USACE Floodplain Management Assessment of the upper Mississippi River Basin to development of an UMR&T systems plan. Place this assessment under the Mississippi River Commission, operating under the USACE.

10.5: Following completion of the survey, seek authorization from the Congress to establish the UMR&T project.

10.6: DOI should complete an ecological needs investigation of the upper Mississippi River Basin and provide a report to the Administration within 30 months.

10.7: Provide an early report in the USACE Upper Mississippi River—Illinois Waterway Navigation Study of environmental enhancement opportunities in the upper Mississippi River.

10.8: Provide a report on the ecological effects of relocating navigation pool control points under the USACE Navigation Study.

10.9: The Administration Interagency Ecosystem Management Task Force should select an Ecosystem Management Demonstration Project within the upper Mississippi River Basin, and establish a cross-agency ecosystem management team under DOI to develop plans and budgets for the project.

11.1: The USGS should establish a federal clearinghouse for data gathered during preparation of the Review Committee report.

11.2: FEMA should investigate the costs and feasibility of completing a national inventory of floodprone structures.

11.3: The USACE, NWS and USGS, with other collaborators, should continue development of basin-wide hydrologic, hydraulic, and hydrometerologic models for the upper Mississippi River system.

11.4: The Hydrology Subcommittee of the Federal Interagency Advisory Committee on Water Data should review the current standards for computing discharge-frequency relationships in light of observations from the 1993 flood and other recent large floods in the upper Mississippi River Basin.

11.5: The Administration should support the USGS in development and acquisition of detailed digital topographic data and other land characteristics for use in floodplain management and other water resources management activities. Existing DOD technologies should be leveraged to assist in the acquisition of these data.

11.6: The Administration should direct that scientific research be conducted to identify state-of-the-art techniques or applications for estimating and assessing environmental and social impacts.

11.7: The USACE and USDA, in collaboration with the DOI, should evaluate the effect of natural upland storage and floodplain storage in such areas as wetlands and forested wetlands on main stem flooding.

APPENDIX D

RECOMMENDATIONS PROPOSED BY THE SAST REPORT*

2.1: Establish an interagency clearinghouse for data relevant to monitoring and managing the Upper Mississippi River Basin. This clearinghouse should be part of the National Information Infrastructure (NII) and should be well integrated with the various components of the NII, such as the National Spatial Data Infrastructure (NSDI), the proposed National Biological Information Infrastructure (NBII), and other thematic data infrastructures as may be established.

2.2: Establish a permanent interdisciplinary team of senior scientists from the Federal and state governments to coordinate data acquisition, analysis and research to support management and policy decision making for the Upper Mississippi River Basin. Membership on this team should be for a limited term (3 to 5 years) and members should rotate.

2.3: Government and nongovernment organizations should migrate their data to standard transfer formats and structures as the formats and structures become available.

2.4: Organizations producing data sets, even if the data sets are specialized or limited, should list those data sets in the clearinghouse for informational purposes. In times of emergency, these data sets can have broad value.

* Excerpted from *Science for Floodplain Management into the 21st Century.* Preliminary Report of the Scientific Assessment and Strategy Team to the Administration Floodplain Management Task Force, 1994.

2.5: Where important data are not within the purview of the Federal government or cannot be maintained or distributed by government organizations, pointers to appropriate nongovernment data could be used in the clearinghouse. The data could be acquired by users from the producer on an as needed basis.

2.6: Identify specific high resolution data needs, develop standards and specifications for the data and develop cooperative arrangements with organizations that would have the capability and logical need to acquire those high resolution data for other purposes. These organizations would often be state and local governments.

2.7: Develop and maintain a framework data set for the river basin. The framework should form the basis upon which other data could be related and registered. This framework data set should contain physical spatial data, biological/ecological data, social data and economic data to which other data sets could be attached, registered, compared, and analyzed. In addition to standard recommended framework data, such data as river miles, levee locations, and physical features such as bluff location, should be included.

2.8: Establish an interagency interdisciplinary group of scientists to continually refine and develop relevant data and techniques to assist in flood mitigation and response. The group should be coordinated by FEMA, USACE, and USGS and should include agencies such as NWS, TVA, SCS, and others as appropriate. Similar groups should be established for other types of emergencies.

2.9: Provide funding for the SAST to conduct initial analysis and research using the data during FY 1995. Instruct agencies to request funding for continued analysis in their FY 1996 budget request.

2.10: Conduct a crosscut of Federal agency data acquisition, analysis, research, and scientific information distribution activities of benefit to river basin and detailed floodplain management for the Upper Mississippi River Basin (and potentially for the Nation) under the auspices of OSTP and OMB.

2.11: Establish an archive of biological data to be part of the clearinghouse. The National Biological Survey (NBS) should be the lead agency to archive existing biological data sets and to identify and develop basin-wide biological data sets to be used for inventory and modeling in the Upper Mississippi Basin. These objectives coincide with the goals of the NBS' Gap Analysis program as well as the overall mission of the NBS. Current activities by EPA to produce biological data sets for the region should be compatible with the SAST database and EPA should incorporate these data into the clearinghouse.

2.12: Develop critical infrastructure data sets to be incorporated into the SAST database. Data sets should be maintained by the agency responsible for the particular infrastructure. Each responsible agency should become part of the clearinghouse.

2.13: Databases that are crucial to managing the watershed should be upgraded to provide good quality point locations that are suitable for analysis. The tracking system of damage to all infrastructure should be improved to allow for a better reporting mechanism. The system should contain information on type of infrastructure, type of damage, frequency of flood related damage, and historic cost of repair.

2.14: Analysis should be conducted by EPA, USDA and DOI to determine if inundation caused mobilization or other problems with hazardous substances as a result of the 1993 floods.

2.15: Conduct further study to determine how to protect the environment from hazardous materials either by reducing the risk of mobilization, or by neutralizing the effects of mobilization.

2.16: EPA should upgrade the spatial accuracy of the RCRIS, FEDS, PCS and NPDES databases to improve their usefulness for spatial analysis. These data should be maintained current and kept in an EPA node of the clearinghouse to ensure rapid and easy access by users.

2.17: Plans, proposals, and reports that contain useful scientific, engineering, social, or economic data, information and analysis should be listed on a widely used data directory. These documents should be made readily accessible to the user community. Listings should be retrievable by geographic location, topic, organization, and, for plans and proposals, outcome or implementation.

3.1: Federal, state and local agencies should cooperate in programs to: (a) develop hydrologic and hydraulic models to define flood-response characteristics, including groundwater flow and transport of water-quality constituents, of river-floodplains and upland watersheds in the Upper Mississippi River Basin; (b) determine drainage-basin physiographic, land-use, and land-management practices on flood response; and (c) extend and enhance existing observation networks as needed to obtain surface-water, groundwater, water-quality, and meteorological data required for development and application of hydrologic and hydraulic models.

3.2: Federal, state and local agencies should cooperate in programs to: (a) develop up-to-date coordinated estimates of statistical flood frequency-magnitude relations, including flood-elevation profiles and flood-risk delineation maps, at gaged and ungaged sites, for use in floodplain planning, management, and regulation in the Upper Mississippi River Basin; (b) review and evaluate existing and proposed methodologies for statistical flood frequency estimation, at both gaged and ungaged sites, and identify or develop improved methodologies, if necessary; and (c) review and evaluate the adequacy of the existing streamflow-gaging network for defining flood risk in the Upper Mississippi River Basin, and extend and enhance the network, if necessary.

4.1: Develop a regionalization scheme (based on existing or new data, whichever are appropriate), to prepare hydrologic response units (HRU). These HRU's are used for developing broad scale hydrologic models and for evaluating the effectiveness of different potential nonstructural actions in uplands for reducing flooding.

4.2: Accelerate production of soils data of finer resolution than the STATSGO data, e.g., SSURGO data, and include those data as part of the clearinghouse. The SCS should maintain these data for distribution in a generally accepted format that can be easily converted to other generally accepted formats.

4.3: The U.S. Fish and Wildlife Service should complete the National Wetlands Inventory. Agricultural wetlands should be included. This classification would improve the usefulness of the data set for evaluating areas of wetlands and for local and subregional entities to use in planning.

5.1: The USGS, SCS, NBS and USACE should initiate a scientific research program to conduct detailed geomorphic/hydrologic/topographic mapping of the lower Missouri River floodplain and selected upper and middle Mississippi River floodplains to develop an overall geomorphic physical-process model, stratigraphic framework, and geotechnical database. The program would identify floodplain zones of variable flood risk and would include analysis of the age of floodplain zones, sedimentation rates and the record of prehistoric large floods.

6.1: Test for possible cause and effect relationships between observed changes in water level at low and high discharges and factors such as river channel-floodplain modifications, measuring the estimating methods, and natural variation due to different water temperature, sediment loads and changes in land cover.

6.2: Expand depth and velocity hydraulic model simulations to the entire lower Missouri River and determine whether or not river flow regulation has also affected seasonal hydrographic, temperature and turbidity patterns.

6.3: Identify the types and analyze the distribution, acreage and ecology of floodplain wetlands created, destroyed or modified by the flood of 1993. Newly created wetlands should be reexamined periodically by remote and ground surveys over the next decade to document their morphometric changes, longevity and ecological value.

6.4: Further examine historic temperature and river stage records to evaluate the temperature-river stage coupling relationship for the Mississippi, Missouri and Illinois Rivers.

6.5: Resource agencies should capitalize on the opportunity provided by the flood of 1993 to further evaluate use of newly-created aquatic and wetland habitats by a diversity of river-floodplain plant and animal communities. This information should be used to identify the habitats that offer the greatest potential acquisition benefits for restoration programs.

6.6: Develop an inventory of species, communities and habitats and understanding of their functional interrelationships.

6.6a. Analyze information on channel geometry and channel cross-sections for available historical river surveys on the Missouri and Mississippi Rivers.

6.6b. Document changes in land cover and land use and changes in the composition, distribution and abundance of native and introduced flora and fauna over the last 100 years in the uplands and floodplains.

6.6c. Quantify distribution and area of floodplain wetlands at various flood stages under various levee scenarios.

6.6d. Identify sources, sinks and transport mechanisms of nutrients, organic matter and human pollutants in the river-floodplains and their major tributaries.

6.6e. Quantify the agricultural potential of lands in the floodplains with respect to other environmental benefits.

6.6f. Determine the characteristics of an appropriate flow regime (flood pulse and low flow) for maintaining a semblance of pre-control aquatic ecosystems.

6.7: Develop a basin-wide ecological model.

6.7a. Define effects of sediment in runoff to streams and upland impoundments in relation to land management and stream riparian corridors.

6.7b. Determine effects of stream sinuosity on stream flows and flood stages.

6.7c. Evaluate floodplain wetlands with respect to flood stream hydraulics.

6.7d. Evaluate effects of riparian vegetation and other land covers on conveyance and stage of flood waters and the role of riparian corridors as energy dissipaters during flooding.

6.7e. Evaluate effects of upland restoration on stream hydraulics at both flood and low flows.

6.7f. Determine if, and how much, upland and floodplain wetlands influence flood peaks.

6.7g. Define environmental and ecological requirements of important aquatic and terrestrial river-floodplain species.

6.7h. Characterize the important compositional, structural and functional interrelationships among plant and animal communities of the major Upper Mississippi River Basin rivers and their floodplains.

6.8: Establish ecological restoration sites on mainstem and tributary floodplains in the uplands.

6.8a. Identify suitable locations for the sites.

6.8b. Develop criteria to measure the effectiveness of ecological restoration sites.

6.8c. Establish a monitoring system to determine how well the restoration sites are meeting the criteria.

6.8d. Modify the restoration and ecological management practices as new understanding is gained.

7.1: Acquire higher resolution topographic data for areas of the basin where low relief potholes exist to allow more accurate determination of pothole storage volumes.

7.2: Conduct field trials and demonstration projects to determine the effect of various land management practices on flood dynamics, sedimentation and soil conservation, agriculture and habitat restoration. These studies should be conducted in a variety of physiographic regions.

7.3: Model additional watersheds within the basin to better estimate the flood reduction effects of upland treatment measures for other terrain types.

8.1: The USACE in cooperation with USGS and SCS should conduct a detailed historical analysis of levee breaching to document specific levee locations and causes of high failure rates. This study should include geotechnical data and new field studies of hydraulic and geomorphologic factors that directly affect levee erosion and failure.

8.2: On the basis of detailed floodplain mapping and historical levee evaluation, the USACE in cooperation with USGS and SCS should identify alternative alignments for levees with high failure rates.

8.3: On the basis of new hydraulic modeling of design floods, USACE and USGS should develop new levee designs that permit passive flooding of protected areas during major flood events to reduce levee damage and floodplain erosion and sedimentation associated with levee breaches under high head conditions.

8.4: The levee data set should be maintained by the USACE in coordination and cooperation with the SCS and the states. Currency should be maintained on the levee data set. It should be part of the clearinghouse and access should be available to interested parties through the clearinghouse.

8.5: Since the Manning's "n" value is not well understood for shrubs and especially not for deep flows through trees, additional research should be undertaken or continued to determine proper "n" values for use with shrubs and trees during deep flooding. The relationship between various depths of flow and Manning's "n" values should be determined for urban and suburban land covers as well.

8.6: Develop a standard set of cross-sections that can be updated by local, state, and Federal input for the modeling of long river reaches. The availability of a standard set would reduce the cost of modeling and facilitate the development and calibration of a basin-wide model that could be used for planning design, operations and forecasting if necessary. The standard cross-sections must be updated when there are changes in floodplain morphology and must be available through the clearinghouse.

8.7: For the short term, develop a model capable of handling 1-dimensional and 2-dimensional modeling simultaneously such that two-dimensional segments can be set into one dimensional researches, or develop full two-dimensional modeling capabilities for the entire river reaches with the capability to model levee breaches, river junctions, and areas with critical infrastructure to determine flow directions and velocities. For the long term, extend the two dimensional capabilities to the entire floodplain.

8.8: Acquire digital elevation data of sufficient vertical accuracy and horizontal resolution to support development of a two-dimensional hydraulic model of the floodplains of the upper Mississippi, lower Missouri and Illinois Rivers.

8.9: Continue investigating new methods for acquiring topographic data and their associated costs by testing the various technologies to determine which is most technically and economically effective.

8.10: The USGS in cooperation with USDA, EPA, and the states and tribal governments should acquire detailed land use data to support modeling on the floodplain.

9.1: Agricultural and other economics data should be included as part of the clearinghouse to ensure their availability to people conducting analysis and management in the river basin. These data should be maintained by the appropriate producing agencies.

9.2: The data, information and analysis resulting from the USACE floodplain management assessment studies and the SCS studies to determine average annual flood damages and costs and benefits of the PL 83-566 projects should be incorporated into the clearinghouse for the Upper Mississippi River Basin.

9.3: The Economic Research Service of USDA, FEMA, USACE and other Federal agencies, and states should cooperate to acquire data at sufficiently fine resolution to conduct economic analyses to determine the economic viability of various structural and nonstructural alternatives for controlling or reducing the effect of flooding. These data should be maintained as part of the clearinghouse.

9.4: The USGS in cooperation with USDA, USACE, DOC and FEMA should develop and implement a scientifically based statistical method that can be used as an input to risk assessments and cost effectiveness analyses.

10.1: The Upper Mississippi River Basin must be managed as an integrated system.

11.1: Produce a baseline data set for the Upper Mississippi River Basin.

11.2: Establish a monitoring program to identify changes to the Upper Mississippi River Basin system. This program should link to and integrate ongoing monitoring programs for the physical, ecological and socioeconomic sectors of the environment.

11.3: Conduct initial and ongoing scientific analysis of the Upper Mississippi River Basin system. This analysis should build and expand on the work initiated by the SAST. The analysis should be conducted by all levels of government and relevant non government organizations. The analysis should be conducted to improve understanding of how the various parts of the system interrelate and to provide new understanding to policy and management decision makers.

11.4: Develop ecologic and hydrologic models of the river basin and advanced hydraulic models of the floodways of the main stems and major tributaries.

11.5a: Appropriate agencies should manage and maintain the data that are under their purview to standards and specifications that allow intercomparability and interchangeability.

11.5b: Establish a distributed clearinghouse for data and information. This clearinghouse should be part of the National Information Infrastructure and meet the specifications of the Federal Geographic Data Committee.

11.6: Establish a coordinating body of scientists to review programs of scientific research and data collection for various levels of government and advise Federal, state, tribal, and local governments and nongovernment organizations on the scientific activities that are and should be conducted to support the management of the Upper Mississippi River Basin. This interdisciplinary body of scientists would serve as principal scientific advisors to the River Basin Commissions and the Water Resources Council. Additional interdisciplinary scientific bodies should be established to meet specific goals of management and disaster response in the Upper Mississippi River Basin.

APPENDIX E

Recommendations Proposed by House Document 465 in 1966*

Summary of H.D. 465 Recommendations		Implementation Progress[‡]		
		1976	1979	1986
A. To improve basic knowledge about flood hazard:				
1. A three-stage program of delimiting hazards should be initiated by the Corps of Engineers, the Geological Survey and other competent agencies.		B	B	A
2. A uniform technique of determining flood frequency should be developed by a panel of the Water Resources Council.		A	A	A
3. A new national program for collecting more useful flood damage data should be launched by the interested agencies, including a continuing record and appraisals in census years.		C	C	C
4. Research on (1) floodplain occupancy and (2) urban hydrology should be sponsored by the Department of Housing and Urban Development, the Department of Agriculture and the Geological Survey.	(1)	C	C	B
	(2)	B	B	B

* From *Floodplain Management in the United States: An Assessment Report,* 1992, L.R. Johnston & Associates, prepared for the Federal Interagency Floodplain Management Task Force.

‡ From status reports in *A Unified National Program for Floodplain Management:* A = Largely Implemented; B = Some Progress (often legislated, but not implemented); C = Little or nothing accomplished.

	1976	1979	1986
B. To coordinate and plan new developments on the floodplain:			
5. The Federal Water Resources Council should specify criteria for using flood information and should encourage States to deal with coordination of floodplain planning, and with floodplain regulation.	B	B	B
6. Under the following Federal programs, steps should be taken to assure that State and local planning takes proper and consistent account of flood hazard: Federal mortgage insurance Comprehensive local planning assistance Urban transport planning Recreation open space and development planning Urban open space acquisition Urban renewal Sewer and water facilities (Many of the necessary coordinating actions were accomplished during final preparation of H.D. 465.)			
7. Action should be taken by the Office of Emergency Planning, the Small Business Administration and other agencies to support consideration of relocation and floodproofing as alternatives to repetitive reconstruction.	B	B	B
8. An Executive Order should be issued directing Federal agencies to consider flood hazard in locating new Federal installations and in disposing of Federal land.	A	A	A
C. To provide technical services to managers of floodplain property:			
9. Programs to collect, prepare, and disseminate information and to provide limited assistance and advice on alternate methods of reducing flood losses, including floodplain regulation and floodproofing, should be undertaken by the Corps of Engineers in close coordination with the Department of Agriculture.	A	A	A

	1976	1979	1986
10. An improved national system for flood forecasting should be developed by the Environmental Science Services Administration as part of a disaster warning system.	B	B	B
D. To move toward a practical national program for flood insurance:			
11. A five-stage study of the feasibility of insurance under various conditions should be carried forward by the Department of Housing and Urban Development.	A	A	A
E. To adjust Federal flood control policy to sound criteria and changing needs:			
12. Survey authorization procedure and instructions should be broadened in concept.	A	A	A
13. Cost-sharing requirements for federally assisted projects should be modified to provide more suitable contributions by State and local groups.	B	B	B
14. Flood project benefits should be reported in the future so as to distinguish protection of existing improvements from development of new property.	A	A	A
15. Authority should be given by the Congress to include land acquisition as part of flood control plans.	B	B	B
16. Loan authority for local contributions to flood control projects should be broadened by the Congress.	C	C	B

Sources: U.S. Water Resources Council. *A Unified National Program for Flood Plain Management.* Washington, D.C.: U.S. Water Resources Council, 1976 and 1979; Federal Interagency Floodplain Management Task Force. *A Unified National Program for Floodplain Management.* Washington, D.C.: Federal Emergency Management Agency, 1986.

APPENDIX F

THE 1994 UNIFIED NATIONAL PROGRAM ACTION AGENDA FOR 1995-2025*

Objective	**Completion Date**
Goal-Setting and Monitoring	
a. Devise a mechanism for setting, monitoring and revising national goals.	1995
b. Hold a national forum on "Floodplain Management for the First Quarter of the 21st Century," to discuss and modify the mechanism as needed.	1996
c. Institutionalize the mechanism through legal, legislative or administrative measures.	1997
Mitigation of Risk	
a. For all metropolitan floodplains, complete an inventory of	
• all existing structures	1996
• all natural resources	2000
b. For all nonmetropolitan floodplains,	
• inventory all existing structures	2000
• identify areas with high potential for development	2000
• inventory all natural resources	2005
c. Mitigate the risk of flood damage for at least half the Nation's highest-risk floodplain structures.	2020
d. Reduce, by at least half, the risk of degradation of the most important natural resources of the Nation's floodplains.	2020

* From *A Unified National Program for Floodplain Management,* 1994, p. 33. Federal Interagency Floodplain Management Task Force, Washington, D.C.

Objective	Completion Date
Public Awareness	
a. Develop a simple concept and definition of floodplain management that will improve public understanding and support.	1996
b. Lay out a leadership strategy to encourage initiative and acceptance of responsibility.	1996
c. Establish new incentives that give credit for integrating different floodplain management programs, strategies and tools.	1996
d. Devise a national strategy to foster public understanding that mitigating action is required when floodplain development potentially damages public or private property or natural resources.	1997
Professional Capability	
a. Make available enhanced training, especially that which takes a comprehensive view of floodplain management.	1996
b. Establish in-house, professional floodplain management capability in all states and in all metropolitan areas.	1998
c. Provide professional floodplain management services to nonmetropolitan areas.	2000
d. Establish professional standards for floodplain management expertise.	2000

APPENDIX G

RECOMMENDATIONS OF THE 1989 ACTION AGENDA FOR MANAGING THE NATION'S FLOODPLAINS*

We recommend six groups of actions that should be taken by the Interagency Task Force or by other Federal agencies in close collaboration with State and Local organizations.

1. Integrate flood loss vulnerability and protection of floodplain natural values into broader state and community development and resource management processes.

   1.1. To promote integrated planning and management of appropriate hydrologic units, many of which encompass multiple local and state jurisdictions, the Interagency Task Force on Floodplain Management should vigorously foster the preparation of State floodplain management plans involving both public and private interests and, where appropriate, interstate agreements for preparation of basin plans. Such plans should consider and balance measures to preserve and enhance the ecological integrity of hydrologic units with measures to meet social needs.

   1.2. Because comprehensive floodplain management programs provide a means for balancing economic development, flood-loss

* From *Action Agenda for Managing the Nation's Floodplains,* prepared by the National Review Committee, October 17, 1989. Published in *Floodplain Management in the United States: An Assessment Report*, 1992, L. R. Johnston & Associates, prepared for the Federal Interagency Floodplain Management Task Force, Appendix F).

reduction, environmental protection and other community goals, along with means of integrating stormwater quality and quantity objectives with upland and floodplain land uses, sections 1361 and 1315 of the National Flood Insurance Act should be administered so as to require preparation of comprehensive floodplain management plans that complement the two national goals as a condition for continued participation in the National Flood Insurance Program. (Several of our members regard this requirement as impractical because many local governments lack the resources necessary to meet it.)

1.3. As a further incentive for the preparation of such plans, the Interagency Task Force should draft and recommend an Executive Order requiring that new Federal investments, regulation, grants-in-aid and other floodplain actions be consistent with State and Local floodplain management plans insofar as they conform to Federal standards.

1.4. To assist in preparing comprehensive floodplain management plans, the Interagency Task Force should seek to coordinate Federal programs and to foster model plans, demonstration projects and research to improve planning methods and techniques.

2. Improve the data base for floodplain management.

2.1. In jurisdictions expected to experience rapid rates of urban growth in upstream drainage areas, the floodplains should be re-mapped to take into account hydrologic conditions associated with full development of the drainage areas under existing land-use plans and policies of relevant jurisdictions, with a view to curbing increased stormwater runoff.

2.2. A cooperative and jointly funded program should be established by the National Science Foundation and the interested Federal agencies to develop methods for mapping, regulating, and identifying natural values in areas with special flood hazards including: (1) alluvial fans; (2) fluctuating lake levels; (3) ice jams; (4) moveable stream channels; (5) land subsidence; (6) storm drainage overflow and backup; and (7) mud flows, and to develop methods for measuring the flood storage capacity of river reaches.

2.3. The Interagency Task Force should formulate an accurate and affordable national system for gathering flood loss data meeting the needs of policy makers and floodplain mangers.

2.4. The National Science Foundation should be requested to consider funding research to examine, in a selected sample of communities, the full benefits and costs, both public and private, of floodplain occupancy and associated floodplain management measures, having due regard for national productivity, the impacts on natural values, and the equitable distribution of costs and benefits.

3. Give weight to local conditions

3.1. Because uniform national prescription standards for the preservation, use and development of floodplains and other hazard areas sometimes create the potential for inefficient allocation of resources and for social inequities, the Federal agencies should examine the practicability of using performance standards, implemented through local watershed and floodplain management programs, but should not promote any slackening of limits on permissible vulnerability.

3.2. The Federal Insurance Administration should adopt and implement a community rating system to encourage communities to adopt flood hazard mitigation measures particularly suited to their local circumstance. Such a system should recognize the need to reconcile loss reduction, public safety and environmental objectives.

4. Minimization of Conflicts among Federal Programs

4.1. The Office of Management and Budget should establish an independent task force to further review the Status report's findings, and recommend those changes in the Federal structure and delegated legislative authority needed to insure execution of a sound Unified National Program for Floodplain Management.

5. Reducing Vulnerability of Existing Buildings

5.1. Because the vast majority of buildings and infrastructure presently exposed to flood damage will not be protected fully from flood waters by structural projects or nonstructural programs, other approaches are needed at both Federal and State levels. As a first step in addressing that problem, the Interagency Task Force should draft and recommend an Executive Order charging all Federal agencies with the preparation of assessments of the vulnerability to flooding of a sample of Federal facilities and those State and Local facilities constructed wholly or in part with Federal aid. The report should identify the facilities' expected average annual damages, estimate the costs of various protection measures, and extrapolate conclusions on the total Federal investment subject to flood damage. The report should be submitted to the President and the Congress with recommendations on appropriate programs to protect Federal facilities.

5.2. As an aid to coordination of those activities, the Interagency Task Force should report which agencies are undertaking nonstructural damage reduction activities and their funding levels.

5.3. Two approaches, in particular, deserve greater attention as viable damage reduction measures: flood preparedness and retrofitting (floodproofing). The National Science Foundation should be encouraged to fund research on the techniques, benefits and costs of these approaches to identify their utility and impediments to their implementation.

6. Improvement in Professional Skills and Public Education

6.1. Inasmuch as the lack of personnel in Federal agencies and in State and Local government who are trained in the interdisciplinary field of floodplain management is an important constraint on progress in the implementation of comprehensive floodplain management, the Interagency Task Force should develop training programs and conduct regional training, at an affordable rate, of appropriate governmental personnel.

6.2. Recognizing that comprehensive floodplain management programs will be more successfully implemented if they are understood and supported by the general public, the Interagency Task Force and its member agencies should continue, expand and evaluate efforts to inform and educate the public about the nature of flood hazards, the natural values of floodplains and the various strategies and tools available for comprehensive floodplain management.

APPENDIX H

Flood Risk Management and the American River Basin: A Summary*

There is no doubt that Sacramento, California, and the surrounding metropolitan area face a significant flood risk from both the Sacramento River and the American River, which converge at the city's doorstep. More than 400,000 people and $37 billion worth of damageable property are vulnerable to flooding in the Sacramento area, including most of the city's central business district and the State Capitol complex. Although there is consensus that action is needed to reduce the level of risk while allowing reasonable use of floodplains, agreement on the appropriate target level and the approach to achieve it has eluded national, state, regional and local decisionmakers.

In 1991 the Sacramento District of the U.S. Army Corps of Engineers[1] completed a study, the *American River Watershed Investigation* (ARWI), that reviewed the American River's contribution to the area's flood hazard, considered a range of flood control measures, and recommended a preferred flood control strategy (USACE, Sacramento District, 1991). The effort met significant criticism, some of it highly technical and some of it political. As a result, Congress directed the Sacramento District to reevaluate its analysis and gather additional input. In response to this

* Reprinted from *Flood Risk Management and the American River Basin.* National Academy of Sciences, 1995, pgs 1-12. References to chapters in this summary refer to those in that book.

congressional directive, additional study and planning have been done by local, state, and federal interests. These efforts–which continue as this report goes to press–have yielded a more comprehensive picture of the flood risk and a broader array of possible flood risk reduction alternatives. The information available today is more comprehensive and more detailed than that available in 1991.

But the fundamental dilemma remains unresolved: how do we balance the potential benefits, impacts, costs, and trade-offs associated with the identified alternatives and select the best management plan for the basin and its residents? This final decision lies not in the realm of science and engineering, but in the arena of public decisionmaking. It requires participants to set aside differences and seek commonalities. It requires weighing competing values. In the end, it will require leadership from local governments, the state of California, the U.S. Army Corps of Engineers, and Congress, as well as a sincere effort by the region's interest groups to agree among themselves about how to respond to the flood hazard.

THE COMMITTEE'S CHARGE

At the same time that Congress asked for a reevaluation of the potential flood control alternatives available to the Sacramento area, the nature of some of the criticisms caused members of Congress to seek an outside body to review the technical soundness of the analyses and related policy questions. Congress directed the Secretary of the Army to ask the National Academy of Engineering to form a special committee, the Committee of Flood Control Alternatives in the American River Basin, to review the 1991 ARWI, with attention to the contingency assumptions, hydrologic methods, and other engineering analyses used to support the seven flood control options presented. Significantly, the committee was not asked to recommend a preferred alternative; instead, it was asked to evaluate the scientific and engineering knowledge base on which the selection of a final strategy will ultimately be based. The committee also was asked to take a step back from the often acrimonious debate that has surrounded the American River planning process and provide insights of value to other regions in the nation that face similar problems–other areas where cities have grown in flood-prone areas and now face significant flood risks. The massive Midwest floods of 1993 and significant regional floods in 1994 and 1995 are reminders of how serious this issue is for the nation.

The committee's charge contains an inherent dilemma. Because there was great controversy surrounding the 1991 ARWI, the committee was asked specifically to review that document. But the controversy surrounding that document was so great that Congress simultaneously asked the Sacramento District to revise it. As a result, while the committee was gathering information for its analysis, efforts to improve the 1991 report were being made by the Sacramento District, the Sacramento Area Flood Control Agency (SAFCA), the Reclamation Board of the State of California, and the State Department of Water Resources, among others.

Thus in this report the committee comments on the data, analysis and methodologies used in the 1991 ARWI where they are still germane. In addition, where possible, it reviews the new analyses and methodologies being used to reevaluate Sacramento's flood risks and assess alternative flood risk management strategies. This has proven to be a difficult task because the committee's study, the Sacramento District's ongoing efforts, and parallel work through SAFCA continue to move along in near synchrony. Also, at this point there is little written documentation of the new work.

The majority of the information concerning the ongoing work was received informally. The committee spoke at length with technical staff from both the federal and the state agencies and tried to understand what methodologies and data were being employed in the current analysis. The committee also heard from a variety of interest groups. In 1994, a new document, *Alternatives Report: American River Watershed, California* (USACE, Sacramento District, 1994a), reached the committee in time to be considered, but this interim report lacked detail. For a true reevaluation of the Sacramento District's technical analysis, Congress might wish to request a review of the upcoming *Draft Supplemental Information Report and Environmental Documentation*, expected to be available in the summer of 1995, because that document will update the 1991 ARWI in detail.

THREE PREMISES

As the committee conducted this review of the Sacramento District's planning for flood control in the American River basin, it became clear that the members shared certain premises (i.e., assumptions believed to be true on the basis of experience and expertise) that influenced their thinking. These premises are: (1) the belief that alleviation of Sacramento's flood risk is critical; (2) the belief that decision making in the American River basin should not stand in isolation, but should be seen in light of national policy that stresses the use of multiple strategies to respond to flood hazards; and (3) the belief that technical matters cannot be neatly separated from policy judgments. These three premises are introduced here to provide a context for understanding the scope of the committee's review and the nature of this report's conclusions and recommendation. These introductory ideas are followed by brief overviews of the chapters of the report.

Alleviation of Sacramento's flood risk is critical

Actions to alleviate Sacramento's ongoing flood risk are urgently needed. The flood-prone development in Sacramento is intense and of high value. This situation occurred in response to historical influences and cannot now be reversed. Nearly 10 years have elapsed since the city's existing flood defenses were clearly proven to be inadequate in the flood of February 1986. Although the careful analyses necessary to support decisionmaking take time, especially if the process is to allow adequate public participation, there comes a point where talk must turn into action. For a variety of reasons, the public decisionmaking process has been blocked from reaching consensus on a feasible course of action to provide the Sacramento area with a higher level of security. Paradoxically, efforts to enhance protection for the largely undeveloped floodplain of the Natomas Basin have progressed further in Congress and locally than proposals affecting developed areas, including downtown Sacramento and the State Capitol complex. Ultimately, California and the nation need to reexamine their approaches to public decisionmaking. Widespread involvement by stakeholders and careful consideration of all options is of course necessary. But delay per se can be counterproductive, costly and potentially dangerous.

Sacramento, California, is a city that grew literally at the edge of the American River and it has been plagued by recurring floods as a result. More than 400,000 people and $37 billion worth of damageable property are vulnerable to flooding in the area, including most of the city's central business district and the State Capitol complex. (Robert Childs, U.S. Army Corps of Engineers.)

National flood policies urge multiple adjustments to hazard

Flood control in California cannot be treated in isolation but must be treated as a part of a complex system of water control and use that has evolved over a long period under the auspices of many government agencies and in response to significant pressures. As the committee approached the task of assessing flood risk along the lower American River, it was aware of recent laws and policy reviews that reflect a broadening of our nation's response to floods over the past quarter-century. For decades, the predominant response to flood risk was to build large flood control projects–dams, reservoirs, levees, diversion channels–to store and restrain floodwaters. The adoption of the National Flood Insurance Act of 1968 marked a watershed in national policy on flood hazards because it established nonstructural measures–flood insurance, floodplain management, and selective acquisition–as mainstays of national flood policy. Additional nonstructural measures in widespread use today include flood forecasting, evacuation planning, public education, and floodproofing of individual commercial and residential structures located in floodplains.

More recently, the 1994 Unified National Program for Floodplain Management (FEMA, 1994) also called for a blend of strategies, from structural approaches to modify flooding to restoration of floodplains. A major evaluation of the Midwest floods of 1993 prepared by the Interagency Floodplain Management Review Committee (IFMRC, 1994) at the direction of the White House, which calls for "shared responsibility" among all levels of government and private interests in responding to flood hazards, also strongly supports the use of nonstructural measures such as relocation of structures out of floodplains and restoration or wetlands, where feasible.

As noted by the National Review Committee (1989):

> The present status of floodplain management does not encourage complacency. The record is mixed. There are encouraging trends, as with the number of communities having some form of floodplain regulations, but the rising toll of average annual flood losses has not been stopped or reversed. Some activities look more productive on paper than on the ground or in the real vulnerability of people. On balance, progress has been far short of what is desirable or possible, or what was envisaged at times when the current policies and activities were initiated.

Thus planners and decisionmakers should proceed with caution. No single technical or institutional "fix" is likely to be an adequate response to the lower American River flood hazard. Responsible federal, state, regional, and local officials must seek to identify a combination of policies and measures that will maximize flood reduction benefits while minimizing economic and environmental costs.

Technical assessment includes policy judgments

The charge to the committee was based on the premise that many of the criticisms of the 1991 ARWI were matters of technical dispute and that a technical judgment could be rendered about the merits of the critics' comments. Representatives of USACE,

SAFCA, environmental groups and Congress at different times emphasized that the committee should try to settle the technical debate, in order to let the political process make the public policy choices about the acceptable risk at Sacramento, including Natomas. However, the planning and design of a flood control program although requiring complex modeling, engineering and data manipulation, do not divide neatly into two parts, technical analysis and policy decisions. For example, even the most apparently technical computational concerns, such as what to assume about the likely coincidence of peak flows at the confluence of two rivers or about use of surcharge space, are based on a policy viewpoint about the acceptable risk of modeling error.

IDENTIFICATION AND EVALUATION OF ALTERNATIVES

As the committee conducted its review of the American River planning process, it noted that perhaps the most critical step in the development of a flood control strategy is the selection of alternatives for detailed analysis. The 1991 ARWI presented various alternative approaches to providing flood control for the American River basin, addressing level of protection provided, costs, expected benefits and environmental impacts. The report was controversial, and some criticisms were based on the perceived failure of the Sacramento District to consider and evaluate range of effective alternatives, such as modification of the operation of Folsom Dam coupled with improvements in outlet capacity. In considering the issue of alternative flood control plans in both the 1991 ARWI and a more recent document, the 1994 Alternatives Report, the committee focused on four issues: (1) use of Folsom Reservoir; (2) the question of gates should a dam be built at the Auburn site; (3) the viability of the Deer Creek alternative; and (4) the adequacy of the nonstructural measures presented.

As detailed in chapter 3, the committee concludes that the original 1991 ARWI was reasonably complete, especially as supplemented by the 1994 Alternatives Report. One concern that arose involved the operating policies employed at Folsom Dam. However, ongoing investigations are now exploring the more dynamic use of Folsom storage capacity. Another concern is the fact that the committee was unable to evaluate how Folsom reoperation was actually considered in the 1994 Alternatives Report, particularly what assumptions were used regarding the initial conditions. These concerns are expected to be addressed in upcoming documents; resolution of these questions should not slow the planning process.

The committee notes that Folsom Reservoir, despite its limitations, is the critical component in the flood-control system for Sacramento. Consequently, it is essential that it be operated as efficiently as possible, and thus the soon-to-be released Folsom Flood Management Plan is critical. It is also important that the operation plan for Folsom evolve as necessary in response to changes in the American River system.

Regarding possible construction of a dry dam at the Auburn site, the committee notes that, should a dam be built, operational gates are essential for dam safety and to provide flexibility in the dam's operation, allowing operators to coordinate with Folsom and other flood control facilities, and to minimize environmental impacts in the upper American River canyon by regulating drawdown.

ENVIRONMENTAL ISSUES

A key issue in the controversy of how to provide flood hazard reduction to the American River basin is how to minimize environmental impacts. Environmental issues were at the heart of many of the disagreements that resulted from the 1991 ARWI. Among the most contentious were the question of the adequacy of the report in assessing potential environmental damage and the uncertainty surrounding impacts of a detention dam in Auburn canyon.

Overall, the committee finds that from an environmental perspective the 1991 ARWI suffered from a lack of scientifically based descriptions of potential impacts and thus did not adequately support the decisionmaking process and help the public weigh the environmental impacts for the range of flood damage reduction alternatives presented. The report understated some environmental impacts, particularly in the upper canyon. The 1994 Alternatives Report, subsequent research and a report from the Lower American River Task Force (SAFCA, 1994b) show significant improvement in understanding impacts, consideration of options, and minimization of impacts.

On the basis of the research to date, the major uncertainty is potential impacts on canyon slopes and vegetation from inundation behind an Auburn detention dam. If such a dam is to be seriously considered, the committee recommends the formation of a multidisciplinary research team to design and carry out a program to reduce this scientific uncertainty and recommend a gate design and operating strategy that could be followed to minimize environmental impacts.

RISK METHODOLOGY

USACE has adopted new risk and uncertainty analysis procedures that are an extension of the traditional paradigm for flood control project planning and community flood protection evaluation. The 1994 Alternatives Report indicates that the Sacramento District's analysis now considers varying degrees of uncertainty in the causes of flooding, such as inflow to Folsom Reservoir, regulated outflow-frequency relationships for Folsom Dam, river stages and levee stability. The methodology computes the risk of flooding due to combinations of hydrologic events, hydrologic parameter uncertainty, uncertainty in reservoir operations, stage-discharge relations and levee performance. USACE traditionally has included safety factors in its design of facilities and the specification of operating policies to address important hydraulic and operational uncertainties in flood control planning calculations; with its new risk and uncertainty analysis methodology, one can investigate the extent to which such safety factors are economically justified.

The committee concludes that the USACE risk and uncertainty procedures are an important initiative. The explicit recognition of modeling uncertainty should result in a better understanding of the uncertainty of flood risk and damage reduction estimates. This change in methodology is important to the American River planning process because the ongoing evaluation of flood control alternatives for the basin is one of the first applications of the methodology. It is almost certainly the most complex application yet attempted by USACE.

As discussed in chapter 4, the new risk and uncertainty procedures, which directly include hydrologic uncertainties in the calculation of average flood risk and the average annual flood damages, tend to inflate those estimates. This tendency can yield benefit-cost calculations more favorable to project justification. The chapter suggests how risk, variability and uncertainty in hydrologic, hydraulic and economic processes should be conceptualized and how the calculations can be organized to avoid introducing such biases while still communicating residual risks and associated uncertainty.

The committee also questions the value of the system reliability index computed by the Sacramento District in its American River study. The 1994 Alternatives Report was found to be particularly confusing because no distinction was made between estimates of flood risk calculated with the traditional level of protection and those calculated with the new risk and uncertainty procedures. Such distinctions are important. USACE needs to develop a consistent scientific methodology and an effective vocabulary for communication of residual flood risks and uncertainties to technical and public audiences.

FLOOD RISK MANAGEMENT BEHIND LEVEES

History shows a close relationship between flood protection and development in flood-prone areas. From the mid-1930s to the late 1960s federally subsidized flood control projects such as levees and upstream storage were the prevalent form of national response, but nevertheless flood losses continued to rise because of continuing development on floodplains. The reasons are many and complex: floodplains can appear to be desirable building locations, and the hazards sometimes are not seen or are unavoidable. Sometimes, development actually is encouraged by federal protection. Once a levee is built to protect development in a floodplain, for instance, it opens the way for additional development, which in turn prompts demands for higher levels of protection. Such development can impose heavy burdens on society. Thus in this era of tightened budgets the question of who pays to support this "flood protection-development spiral" is becoming increasingly important.

One question in the American River basin is whether this flood protection-development spiral is the fate of the Natomas Basin. The Natomas Basin is a flat, marshy lowland of about 55,000 acres near Sacramento that lies entirely within the 100-year floodplains of the American River and the Sacramento River. Today the basin is surrounded by a 41-mile ring of levees and is devoted primarily to agriculture. The basin is now home to 35,000 people, but because of its prime location, it is projected to be a major growth area for new housing and commercial development. Although the existing levees lessen the flood risk to some degree, the Natomas Basin faces significant residual risk. The basin lies below the levels of the American and Sacramento rivers at flood stage and could fill like a bathtub in the event of a flood that breaches or overtops the levees.

According to plans prepared by local authorities, large portions of the basin are poised for development despite the unresolved and perhaps unresolvable issue of its flood hazard. Clearly, the Natomas Basin is well situated in terms of proximity to Sacramento, but it is poorly situated in terms of chronic flood risk. Improvements in the existing flood protection system, including the reoperation of Folsom Dam, levee expansion and other improvements that are in progress or are foreseeable, can help reduce the risk, but significant residual risk will remain. Development within the Natomas Basin thus should be subject to prudent floodplain management requirements under federal, state and local authority. In addition, the public should be informed of the residual flood risks despite the presence of the levee system.

WATER RESOURCES PLANNING AND DECISIONMAKING

The application of the USACE planning process to the search for acceptable flood control for the American River basin has illustrated the need for reforms in how such decisiosn are made, and the committee believes that the lessons of the American River can be transferred to other areas of the nation. Early decisions, such as Congress's direction to limit the project purpose to flood control, that were made ostensibly to lessen controversy and speed the process instead prolonged the debate because public interests desired a wider view. Indeed, many who commented on the 1991 ARWI were critical of its failure to consider any purpose other than flood control as a planning purpose and noted that this single-purpose approach precluded a more integrated approach to planning. In particular, the dispute over the proposed dry dam alternative has stalled the study process. Although progress is being made to mate environmental restoration concerns with improved levee stability and conveyance in the lower American River, in large part because of the work of the Lower American River Task Force facilitated by SAFCA, efforts to resolve disputes over alternatives and impacts in the upper American River have met with little success.

The current decisionmaking situation in the American River basin can be described as a diffusion of separate interests having access to numerous political and legal veto points, making it far easier to stop an activity than to move one forward. Despite some errors and problems with the planning process as implemented in the American River basin, the committee recognizes that USACE to date has been embroiled in larger California water controversies and at times technical complaints have been used as weapons in a policy dispute. The committee believes that in the American River context and similar situations USACE must make its work part of a shared planning process where the local sponsor, other agencies of the federal and state governments, and nongovernmental interests can cooperatively develop the data and models, understandings of risks and tradeoffs, formulation of alternatives and consensus on the most appropriate alternative.

The American River situation is not unusual; USACE has frequently seen its recommendations challenged in recent years and thus needs to find ways to improve the planning process so it works more effectively in the future. Areas open to reform include (1) acceptable damages and the flood insurance program; (2) water project cost sharing; (3) communication of flood risk; (4) water project planning; and (5) water policy and management at the national level.

FINDINGS AND RECOMMENDATIONS

This committee's task was to evaluate the scientific and engineering knowledge on which the selection of a flood hazard reduction strategy for the lower American River will ultimately be based. The committee also endeavored to provide insights on public policies concerning flood hazard management that are of concern to the nation. In line with that dual charge, the committee offers findings and recommendations specific to the USACE planning process as applied to the American River basin, as well as some broader comments on the nature of flood risk assessment and its application nationwide.

The findings and recommendations presented in detail in chapter 7 relate to (1) the identification and evaluation of alternatives; (2) environmental issues; (3) risk methodology; (4) flood risk management behind levees; (5) risk communication; and (6) water resources planning and decisionmaking. Some of the key issues are summarized here, but chapter 7 provides a fuller treatment of the findings and recommendations.

- Overall, the committee finds that the 1991 American River Watershed Investigation, as supplemented by the 1994 Alternatives Report, was reasonably complete in its consideration of structural flood protection measures. Alternative assumptions could have been selected, but nothing of a degree that should call the overall results into question.
- The committee does not and can not judge whether construction of a dry dam at the Auburn site is the best approach to reducing Sacramento's flood risk. However, the committee strongly believes that if a dry dam is built it must contain operational gates to ensure management flexibility, protect public safety and minimize environmental impacts. Environmental concerns are significant, and additional research is needed to understand the potential impacts of a dam on the canyon environment, particularly plant communities and slope stability. Such information could be used to help set operational guidelines so impacts of such a dam could be minimized. In addition, if a dam is built, the

committee believes it should be used as a last line of defense to contain peak flows from extreme floods, thus reducing the frequency of impacts on the canyon.

- The new USACE risk and uncertainty procedures are an innovative and timely development. The explicit recognition of modeling uncertainty should result in a better understanding of the uncertainty of flood risk and damage reduction estimates. However, the committee is concerned about the specific ways in which uncertainty is currently represented and included in the calculation of average flood damages and the residual risk of flooding, and about USACE's ability to communicate information about flood risk and community vulnerability. USACE leadership is encouraged to convene an intra-agency workshop, including outside experts, to review the new risk and uncertainty procedures.
- The determination of the federal interest in construction of water management facilities has always been a complex process affected by many factors, such as societal goals, the nature of the problem to be addressed, and financial constraints. The rationale for federal interest in flood control in the American River basin should be reviewed, and Congress should explicitly address whether federal involvement is warranted on the basis of the presence of widespread national benefits from flood protection or a limited ability of the community to provide its own flood protection. If a federal interest is clear, project construction should be delayed until SAFCA, working with FEMA and private insurers, has a program to require that new development at Natomas and in the city purchase flood insurance at actuarially sound rates for the residual risk. Also, SAFCA should implement a flood hazard mitigation plan, to be part of the area's land use plans, that includes flood risk communication, flood warning systems, evacuation plans to reduce loss of life, highway and other infrastructure designs to facilitate evacuation, and floodproofing and elevation requirements.

The fundamental question in the American River planning process is how to reduce flood risk in the lower American River basin given a decisionmaking arena that includes significant scientific uncertainty and organized opposition to some of the possible risk reduction alternatives. This report discusses the uncertainties that confront floodplain managers and offers recommendations in many areas, including the need for additional research in some areas. But decisionmakers, agency officials and interest groups reading this report should not use calls for additional research as an excuse for not taking action. It is time to select and implement a flood risk reduction strategy for the American River basin. There are still areas where data and information are incomplete, particularly in our understanding of environmental impacts, but that should not forestall the decisionmaking process. The recommendations offered in this report are intended to improve the process, not delay it further.

THE ROLE OF SCIENCE IN THE DECISIONMAKING PROCESS

The issue decisionmakers face is how best to determine and then implement an acceptable flood risk management program for the American River basin. Beyond all the complexities and subtleties, the ultimate question is whether the flood damage reduction offered through a combination of measures not including a dam is acceptable, or whether a new upstream dam is judged to be necessary to reduce risk to an acceptable level. The committee cannot answer that question, in part because detailed technical analyses comparing the alternatives are still being developed (this information is expected in the Sacramento District's forthcoming *Supplemental Information Report*, scheduled to be available in the summer of 1995) and, importantly, because that judgment is beyond the committee's appropriate role. The public should be forewarned that even when the technical analyses are available, there will be no simple technical answer. Scientists and engineers can and should provide careful analyses and interpret the information so it is available to support decisionmakers, and they should be frank about uncertainties and risks. But the decision to be made should ultimately reflect more than technical factors; it should reflect economic considerations and value judgments pertaining to the appropriate use of natural resources, public monies, acceptable levels of risk, and willingness to accept constraints on land use. The final decision on these issues rests with the public and the political officials who represent them.

Reference

1. USACE (U.S. Army Corps of Engineers) as used herein refers to actions taken by the Washington D.C. headquarters of the Corps of Engineers (e.g., agency-wide policies, procedures, etc.) or comments by the headquarters on subordinate activities by subordinate elements such as the Sacramento District. Field activities, reports, work in progress, meetings, etc. by the Sacramento District are identified as the Sacramento District unless and until specifically acted on by USACE.

APPENDIX I

THE FPMA: EXECUTIVE SUMMARY*

The Midwest Flood of 1993 was without precedent in many respects, such as the areal extent and duration of rainfall that led to it, the severity of flooding at many locations, and the institutional response of the nation. The ensuing public attention and reaction generated Congressional authorization and appropriations for the Corps of Engineers to conduct a comprehensive, system-wide study to assess flood control and floodplain management in the areas that were flooded in 1993.

The Floodplain Management Assessment of the Upper Mississippi and Lower Missouri Rivers and their tributaries, or FPMA, was authorized by House Resolution 2423, dated November 3, 1993. Congress provided funds in the Fiscal Year 1994 Energy and Water Development Appropriations Act, which was signed into law as Public Law 103-126.

The authorizing language from Congress and subsequent guidance provided by Headquarters, U.S. Army Corps of Engineers established the following 11 objectives for the conduct of this assessment.

a. Describe resources and project future conditions;
b. Identify desires of local interests;
c. Describe varying outputs from alternative uses of floodplain resources;
d. Describe forces that impact floodplain resources;

* Reprinted from *The Floodplain Management Assessment of The Upper Mississippi River and Lower Missouri Rvers and Tributaries,* US Army Corps of Engineers, June 1995. References to chapters in this summary refer to that report.

e. Array alternative actions;
f. Evaluate and prioritize alternatives based on consultation and coordination through public workshops or similar mechanisms;
g. Prepare a report to document efforts, present conclusions, and recommend subsequent follow-on studies;
h. Identify critical facilities needing added flood protection;
i. Examine differences in Federal cost sharing on the upper and lower Mississippi River system;
j. Evaluate cost effectiveness of alternative flood control projects; and,
k. Recommend improvements to the current flood control system.

The FPMA has attempted to be responsive to these objectives while complementing the work accomplished by many others on related aspects of the floodplain issues.

Probably the most notable work by others is the report commonly referred to as the *Galloway Report*. The Administration's Interagency Floodplain Management Review Committee published the report in June 1994. The committee was formed to take a fresh look at floodplain management and other policies that may have contributed to the severity of flood damages. The recommendations of the report are, as of this writing, under consideration by the Administration. Some of the needed changes in Federal flood insurance and disaster assistance programs identified in the report are already enacted into law. The FPMA has attempted to complement the Galloway Report in those areas where the Corps is uniquely qualified.

The FPMA focuses on a comparison of impacts and costs of implementing a wide array of alternative policies, programs, and structural and nonstructural measures by assuming they had been in place at the time of the 1993 flood. It explores three scenarios of changes in flood insurance, State and local floodplain regulation, flood hazard mitigation and disaster assistance, wetland restoration, and agricultural support policies. The structural alternatives ranged from levees high enough to contain the 1993 event to totally removing the levee system, with several intermediate alternatives. This approach brackets the extremes. An acceptable solution is probably somewhere in between and involves a combination of alternatives. A preliminary examination is made of the hydrologic and hydraulic effects of watershed measures and wetland restoration.

These impact analyses are based on results of systemic hydraulic computer modeling that represents an advancement in the state-of-the-art in flood analysis. This modeling work was initiated by the Corps of Engineers prior to the FPMA, but funds were also budgeted under the FPMA. Work performed for the Assessment contributed to the achievement of the first hydraulic modeling capable of predicting impacts of random changes in floodplain storage parameters (such as when a levee break occurs).

Since the beginning of the assessment in January 1994, Corps of Engineers Headquarters' direction has been to include any conclusions that data collection, hydraulic modeling and impact evaluations could support. The goal has been to identify and evaluate alternative floodplain and flood management measures, including the effects of policy changes and modifications to the current flood damage reduction features in the areas that were flooded in 1993.

The FPMA is also unprecedented because of the high degree of cooperation and teamwork displayed not only by the five Corps of Engineers Districts (St. Paul, Rock Island, St. Louis, Kansas City and Omaha), three Division offices (North Central, Lower Mississippi Valley and Missouri River), and Corps Headquarters, but by the representatives of the Natural Resources Conservation Service (NRCS), the Federal Emergency Management Agency (FEMA), the U.S. Environmental Protection Agency (USEPA), the U.S. Fish and Wildlife Service (USFWS) and the states (namely Illinois, Iowa, Kansas, Minnesota, Missouri, Nebraska and Wisconsin). The contributions of data, participation in workshops, and review and comment on interim study products by these various offices helped give this assessment a breadth of perspective beyond that available from within a single agency. Three series of public meetings were held throughout the study area in June 1994, November 1994 and April 1995. Also, the Plan of Study and "Milestone Packages" were distributed in April 1994, August 1994, September 1994 and January 1995. These efforts were designed to inform and to obtain feedback on strategies, the study process, and data being used for evaluation Adjustments to study tasks during the study period resulted from comments.

The feedback received during coordination of the assessment highlights contrasting views regarding use of the floodplain. Some groups advocate broad floodplain management concepts while others view floodplain management as being inconsistent with flood control and economic development. It is also apparent that flood fighting and associated levee raises are part of a culture of self-reliance held by many of the people who are protected by levees. Many believe that the levees constructed 50 or more years ago were adequate for hydrologic conditions at that time, but that the severity of floods has increased due to actions in the watershed that have increased runoff or because of physical changes in channel or levee capacity. Countering some of these views are the concerns about vulnerable uses of the floodplain which result in high costs of disaster relief following a flood event such as that of 1993 and contribute to adverse impacts on the natural floodplain environment. This assessment does not resolve all the issues or recommend an overall best plan. Rather, it serves as another tool in understanding the relative impacts of various potential actions.

As you review the evaluation results, findings and conclusions please be alert to four areas of caution:

1. The 1993 flood event is used as a base condition to evaluate impacts of changes in policies and structural alternatives, recognizing that the 1993 event is still fresh in everyone's minds and provides a wealth of additional information on the region's vulnerability to extreme flood events. In addition, the 1993 flood was so widespread that an opportunity existed to evaluate varying flooding levels, ranging from a 20-year to over a 500-year event in different areas. Its areal extent and duration make it a unique flood, as every flood is. The FPMA does not provide a complete basis for formulating or recommending projects, because flood frequency analysis and evaluation of life cycle and cumulative benefits and costs must first be accomplished. These were beyond the scope of the FPMA.

2. The Findings and Conclusions of this report are those of the five Districts and three Divisions involved in the FPMA effort.
3. The results of the hydraulic modeling of the various alternatives represent approximate values that are appropriate for an overall assessment. Although further analysis could modify results to some degree, the general trends displayed in this report should remain the same. The unsteady-state modeling used for this assessment addresses the relationship between stage and discharge, but not the relationship between discharge and frequency. The flood discharge-frequency estimates for the Upper Mississippi River are based on a 1970 Federal interagency agreement. There are no current plans for revising these estimates for either the Mississippi or Missouri Rivers based on the 1993 flood or other recent floods. However, there is concern by many, including the Corps of Engineers hydrologists, that those estimates need to be revisited.
4. The data collected were almost exclusively data that were already available, such as the economic damages from the 1993 flood. Much of this data is aggregated at a county level, and is not broken down into floodplain reaches. Although there would be a higher level of confidence with data at a greater level of detail, the data used were suitable for this type of initial systemic evaluation.

SOME OF OUR MORE SIGNIFICANT FINDINGS AND CONCLUSIONS ARE:

• **Structural flood protection performed as designed and prevented significant damages.**

Corps reservoirs performed well, reducing flood water elevations along the main stems of the Upper Mississippi and Lower Missouri Rivers by several feet in most locations. Structural flood protection (urban levees and floodwalls) performed as designed in protecting large urban centers. The Congressional General Accounting Office concluded that "most Corps levees performed as designed and prevented significant damages" (page 11 of report dated February 28, 1995).

- **Approximately 80% of 1993 crop damages region-wide were caused by overly saturated fields, unrelated to overbank flooding.**

At least 50% of the total 1993 flood damages were agricultural and approximately 80% of 1993 crop damages region-wide were caused by overly saturated fields or other factors unrelated to overbank flooding. These losses would not have been affected by changes in floodplain management policies. The best option to address these damages is a rational program of crop damage insurance. Crop insurance reform legislation (Title I of PL 103-354) was enacted late in 1994.

- **Flood damages in urban floodplains with inadequate or no flood protection continue to be a major problem.**

For the 120 counties adjacent to the Upper Mississippi and Lower Missouri Rivers and several of their major tributaries that were the focus of this assessment, urban damages substantially exceeded agricultural losses. Overbank flooding and problems associated with urban drainage and stormwater runoff continue to occur in a number of locations, as confirmed by the 1993 event.

- **No single alternative provides beneficial results throughout the system.**

From a hydraulic evaluation perspective, the FPMA analysis illustrates that no single alternative provides beneficial results throughout the system. Applying a single policy system wide may cause undesirable consequences at some locations. Examination of many factors such as computed peak stages, discharges, flooded area extent and depth within flooded areas is necessary to evaluate how an alternative affects performance of the flood damage reduction system as a whole.

- **It is essential to evaluate hydraulic impacts systemically.**

The importance of evaluating hydraulic impacts systemically is clear from the results of the unsteady-state hydraulic modeling. Changes that affect the timing of flood peaks or the "roughness coefficients" of the floodplain can be as significant as changes in storage volume.

• If all agricultural levees had been successfully raised and strengthened, urban flood protection would have been placed at much greater risk.

If the agricultural levees along the Upper and Middle Mississippi River had been raised and strengthened to prevent overtopping in the 1993 event, the flood stages on the Middle Mississippi would have been an average of about 6 feet higher. Likewise, raising the levees to prevent overtopping on the Missouri River would have increased the stage by an average of 3 to 4 feet, with a maximum of 7.2 feet at Rulo, Nebraska, and 6.9 feet at Waverly, Missouri.

• Flood stage changes resulting from the removal of agricultural levees are highly dependent on subsequent use of the floodplain.

Hydraulic routings, assuming agricultural levees are removed show that, with continued farming in the floodplain, 1993 stages would be reduced an average 2 to 4 feet on the Mississippi River in the St. Louis District (middle Mississippi River). If this area would have returned to natural forested conditions, some of the system would still have shown reductions in stage (up to 2.8 feet), but increases in stages by up to 1.3 feet would also be seen in some locations. In the Kansas City District (lower Missouri River), hydraulic modeling shows changes in stages of -3 to +1 foot for no levees with agricultural use and -3 to +4.5 change with forested floodplains.

• Restoration of floodplain wetlands would have little impact on floods the magnitude of the 1993 event. Agricultural use of the floodplain is appropriate if risk of flooding is understood and accepted.

Converting floodplain agricultural land to natural floodplain vegetation would not reduce stages in some locations but would marginally reduce damage payments in the 1993 Midwest Flood. Agricultural use of the floodplain is appropriate when the residual damage of flooding is understood and accepted within a financially sound program of crop insurance and flood damage reduction measures and when it is compatible with essential natural floodplain functions. Current theories on floodplain function predict that the area needed for an improvement to the natural biota is probably fairly small and that restoration of a series of natural floodplain patches (a string of beads) connected by more restricted river corridors would be practical and beneficial.

• Restoration of upland wetlands would have produced localized flood reduction and other benefits, but little effect on main stem flooding.

Hydraulic modeling of reducing the runoff from the upland watersheds by 5 and 10% predicted average stage decreases of about 0.7 and 1.6 feet, respectively, on the Upper and Middle Mississippi River and about 0.4 and 0.9 feet, respectively, on the Lower Missouri River. However, wetland restoration measures alone would not have achieved this level of runoff reduction for the 1993 event because of the extremely wet antecedent conditions. Restoration of upland wetlands would produce localized flood reduction benefits, but have little effect on mainstem flooding caused by the 1993 event. There are other reasons for why restoration of upland wetlands is very important, such as reduced agricultural exposure to flood damage, water quality, reduced sedimentation, and increased wildlife habitat.

• State and local floodplain zoning can be an effective means of siting critical facilities out of harm's way.

State and local floodplain zoning ordinances and regulations could be most effective in determining the siting of critical facilities that have the potential for releasing toxic or hazardous elements into the environment when flooded.

• More extensive reliance on flood insurance would better assure appropriate responsibility for flood damages.

More extensive reliance on flood insurance would better assure that those who invest, build, and live in the floodplain accept appropriate responsibility for the damages and other losses that result from floods. Expenditures for the 1993 flood through the

National Flood Insurance Program and the Federal Crop Insurance Corporation were less than half of total disaster aid payments.

• **Greater emphasis on flood hazard mitigation actions is justified.**

More emphasis is now being placed on use of flood hazard mitigation measures, especially acquisition of flood-prone structures, as an action that will reduce repeated Federal disaster expenditures and other costs associated with areas of widespread and potentially substantial repetitive flooding.

• **Although there are conflicting public viewpoints on uses of the floodplain, areas of potential agreement exist and need to be pursued.**

Comments heard and read from the public throughout the assessment followed three main themes, with varying degrees of acceptance among the interest groups:

a. Importance of agricultural levees;
b. Need for shifted emphasis to non-structural measures and upland watershed measures; and,
c. Need for greater coordination among agencies responsible for managing the upper Mississippi and lower Missouri Rivers.

• **Better adherence to existing policies is a necessary, immediate, and effective first step for better floodplain management.**

Measures that would reduce damages during future floods that are not dependent upon any revised policies and programs include:

a. Good maintenance of both the existing Federal and non-Federal levee system.
b. State and local interests enforcing land use policies to ensure that new floodplain development does not occur or is constructed to minimize damage potential (raising, floodproofing, etc.)

• **Examples of shifting dependence from disaster aid to flood hazard mitigation and flood insurance are justified.**

A shift from dependence on disaster aid to flood hazard mitigation (floodproofing, elevating or acquiring and relocating out of the floodplain) and flood insurance appears to be occurring. The following examples of measures that warrant further consideration generally follow the Federal philosophy of floodplain management which recognizes that flood damage avoidance should generally be the first defense against flooding, complemented by nonstructural and structural flood protection measures, where appropriate, with public education and flood insurance included as essential components to address the residual risk of flooding:

a. acquisition of structures that are repetitively damaged;
b. more widespread and stricter enforcement of flood insurance requirements for individuals, farmers, businesses and communities (already well underway);
c. enforcing strict consistency in eligibility for the provision of disaster aid;
d. greatly increased emphasis on flood hazard mitigation planning and implementation;
e. assuring that communities and individuals are aware of the degree of risk involved in residing behind a levee or downstream of a dam in a floodplain, especially if less than Standard Project Flood (SPF) level of protection;
f. more effective floodplain management policies and zoning standards at the local level to prevent floodprone development;
g. an expanded boundary for flood risk zones to go beyond designation of "100-year" flood zones for flood insurance;
h. more upland watershed retention measures that will hold or slow rainfall runoff; and,
i. continue structural protection when systemic analysis of impacts and life cycle costs indicate this is the best solution, but with an awareness of the risks associated with induced development.

• **Preparation for even larger floods is needed.**

Floods greater than the 1993 flood catastrophe will happen in the future. It would be prudent to prepare for future floods larger than the 1993 event. When were are properly prepared for catastrophic flood events, smaller floods will be more easily accommodated.

• **Much valuable data such as hydraulic modeling, mapping and data inventories resulted from the assessment study.**

The hydraulic modeling, the gathering and organizing of data and viewpoints and the evaluation of this input for the FPMA should provide an improved understanding of many floodplain management issues. The FPMA has played a part in helping to develop many new "tools" for those involved in making floodplain management decisions. There is now a working unsteady state flow hydraulic model on the Upper Mississippi and Lower Missouri Rivers, digitized land use mapping, an environmental resource inventory, and other products, as listed in chapter 12 of the report.

Through the FPMA analyses, the following efforts are considered to have greatest value in furthering future understanding and enhancing sound floodplain management directions:

a. Inventory and spatial database of levees and other structures in the floodplain;
b. Inventory and GIS database of critical facilities in the floodplain;
c. Additional hydraulic modeling (unsteady state) with more detailed mapping and coverage over portions of the main stem rivers not yet modeled and for the larger tributaries. (A system model, including the Mississippi, Missouri, Illinois, Ohio and Arkansas Rivers is scheduled to be available by the end of Fiscal Year 1996);
d. A real-time, unsteady state hydraulic model and tributary rainfall runoff forecasting models for predicting flood crests in future flood emergencies.
e. Updated hydrology and hydraulics data, including discharge-frequency relationships and water surface profiles.

f. More extensive data and hydraulic modeling of upland watershed areas that have the greatest potential for flood damage reduction;
g. Development and experimental testing of biological response models that are linked to existing hydraulic and hydrologic models;
h. If a system-wide plan for flood damage reduction is desired, economic data must be collected, indicating the specific locations and elevations of damageable property; and,
i. Maintain and update the environmental GIS data base that has been developed in this effort. This data base can serve as an important resource in developing floodplain management strategies for specific reaches and in developing a systemic management plan for natural resources.

As stated earlier, this assessment was limited in its evaluation to comparing impacts of a wide array of policies, programs, and flood damage reduction measures to only a single event, the Midwest Flood of 1993. To develop recommendations or a comprehensive floodplain management plan, either system-wide or for specific reaches, would require a more complete analysis. Such an analysis would ideally include impacts of all possible flood events, life cycle and cumulative costs and benefits, and a more quantitative measurement of impact categories such as environmental, social, human trauma, and cultural. However, this assessment has taken an important step toward achieving a better understanding of the current uses of the floodplain, forces causing those uses, and impacts of various alternative changes in the management of the floodplains.

The bottom line of the assessment was probably best stated in one of the comment letters on the draft report which says, "the assessment validates the view that while structural flood control measures are an important part of an overall floodplain management program, they have limitations and floodplains are best managed through a combination of structural and non-structural measures that fully recognize the inherent risk of occupying flood hazard areas."

INDEX

Items in italics denote figures (f) or tables (t).

G

H

I

J

K

L

M

T

U

V

W

Y

Z